PROTECTING LIFE *on* EARTH

THE STEPHEN BECHTEL FUND

IMPRINT IN ECOLOGY AND THE ENVIRONMENT

The Stephen Bechtel Fund has

established this imprint to promote

understanding and conservation of

our natural environment.

PROTECTING LIFE
on EARTH

An Introduction to the Science
of Conservation

Michael P. Marchetti
Peter B. Moyle

UNIVERSITY OF CALIFORNIA PRESS
Berkeley   Los Angeles   London

The publisher gratefully acknowledges the generous contribution to this book provided by the Stephen Bechtel Fund.

University of California Press, one of the most distinguished university presses in the United States, enriches lives around the world by advancing scholarship in the humanities, social sciences, and natural sciences. Its activities are supported by the UC Press Foundation and by philanthropic contributions from individuals and institutions. For more information, visit www.ucpress.edu.

University of California Press
Berkeley and Los Angeles, California

University of California Press, Ltd.
London, England

Library of Congress Cataloging-in-Publication Data

Marchetti, Michael P.
    Protecting life on Earth : an introduction to the science of
    conservation / Michael P. Marchetti, Peter B. Moyle.
        p. cm.
    Includes bibliographical references and index.
    ISBN 978-0-520-26432-8 (pbk. : alk. paper)
1. Wildlife conservation—Textbooks.   2. Ecology—Textbooks.   I.
Moyle, Peter B. II. Title.
QL82.M358 2010
578.68—dc22   2009050655

Manufactured in China

16   15   14   13   12   11   10
10   9   8   7   6   5   4   3   2   1

The paper used in this publication meets the minimum requirements of ANSI/NISO Z39.48-1992 (R 1997) (*Permanence of Paper*). ∞

If you choose not to be consciously involved in the conservation of forms of life other than your own, you should at least be aware that by doing nothing you are still having an impact on the biota of this planet. The water you drink, the food you eat, the land you live on, and the air you pollute were all obtained at the expense of other creatures. The decisions we make today on how we are going to share these resources will determine which other species will inhabit Earth for the indefinite future.

R. L. Thayer

CONTENTS

INTRODUCTION

The dodo was a large flightless pigeon that once inhabited the remote island of Mauritius. It was clubbed into extinction by sailors in the seventeenth century for food and sport. The dodo is remembered today mainly as a symbol of stupidity: it was too dumb to get out of the way of humans and was therefore wiped out. Unfortunately, most species sharing this "island" planet with us are "dodos." They cannot get out of the way of human activity and will be driven to extinction unless we actively protect them and their habitats. In this book, we show why this last statement is true and also why our own future is tied to that of Earth's diverse animals and plants. Humans and other forms of life on the planet are intertwined and interdependent, so maintaining the biodiversity of living things on this planet is, in fact, in our own best interests.

Why exactly do humans have such an incredibly large influence on other species and the natural world? We are unique among all the animals on the planet in that we survive and reproduce in a wide variety of habitats. Cultural practices have allowed us to colonize nearly every part of the Earth. As a result, the human population has continued to grow exponentially for hundreds of years. Such sustained population growth is unparalleled by any other animal species on the Earth. For most species, their populations grow until they are fully utilizing available resources, such as food and space. At that point, natural mechanisms of regulation, such as disease and starvation, prevent the population from growing further and may cause a population collapse.

However, humans behave differently because we respond to resource scarcity with cultural tools and technologies that halt the natural regulating mechanisms and instead actually increase the availability of resources. We raise our food using intensive, large-scale agriculture, live in dense urban environments, move resources around the globe, and use medical technology to keep us alive longer. This process requires us to continually increase our resource use, and as a result, we accelerate the rate at which we alter the natural environment. Currently, the global human population is large enough and our technologies are potent enough that human-related alterations to the planet are causing unprecedented numbers of species to go extinct. If present trends continue, there will in all likelihood be a crash in the human population as we put demands on our resources beyond what the planet can sustain.

In a very real way, the root cause of the modern biodiversity and conservation crisis

lies in the fact that continued human population growth combined with increased per capita consumption is threatening most life forms on the planet. This book attempts to explain many aspects of this crisis from a biological and scientific point of view. Using accessible and straightforward language, we try to answer many questions about conservation. What are the roots of the crisis, from both a biological and historical perspective? How can understanding evolution and ecology help us to understand the crisis? What does the crisis look like, and how serious is it? Why does this crisis matter? How is society trying to address the crisis from both scientific and legal perspectives? Ultimately, how can we as individuals respond and ultimately help repair the crisis?

In this book, we provide insights into these and many other questions. Our goal is to explain to an intelligent non-scientific audience why conservation of biodiversity is a vitally pressing issue. We live in important and dangerous times, and the only way to see our way through the coming difficulties is to be armed with the power of knowledge. Yet the sheer amount of information we are bombarded with every single day sometimes has the effect of crushing our desire and ability to learn and grow. Much of the human population feels that it is easier to ignore or put aside those things we don't understand. Compounding this problem is the fact that scientists (like us) have a tendency to speak like scientists; they love to convey information using esoteric jargon and revel in the scientific

complexity and uncertainty. Unfortunately, as a result, many people tune out when they hear yet another scientific expert pontificate. We want this book to help bridge the gap between the science of conservation and the people who can and will make the difference: namely, the general public and the people they select to make policy.

In other words, we want to reach out to *you,* the individual who makes hundreds of small but critically important decisions that affect wildlife, biodiversity, and global ecosystems every single day. We want *you* to understand on a straightforward, heartfelt level why conservation is a vitally important topic and how conservation issues touch so many aspects of our daily lives. We want you to come away with reliable information to help inform the challenging environmental decisions we make every day. These are decisions on topics that range from voting on environmentally related issues, to deciding how we as individuals and as a nation spend our money, to determining how we get to work in the morning, to choosing what to put into our mouths when we eat. This book was written to help explain how the science of conservation is at the heart of helping you make informed decisions. You will notice that it has a distinct bias toward using wild vertebrates ("wildlife") as examples. This reflects both the interests of the authors and the fact that we humans have a greater empathy with the creatures that are the most like us.

In addition to targeting a wide audience, we designed this book to be used as a text in a college classroom. It is divided into convenient chapters, each of which addresses a major subject within conservation to make it convenient for instructors to pick and choose topics they wish to cover. The chapters, however, are placed in a logical sequence for learning. We start with a brief introduction to environmental history, followed by introductions to the sciences of evolution and ecology. We then provide a series of chapters on major issues related to conservation of biodiversity, ending with some suggestions as to what you can do to help out the planet.

BOOK ORIGINS

The ideas and examples we use in this text have been field tested over decades of experience in the classroom by both authors. The origins of this book lie in a graduate student exercise in the 1980s by one of the authors (PBM). He gave the students the charge of each writing a chapter for a reader that would explain issues in conservation to a non-science audience. Many drafts later, a version of this early "reader" was used by hundreds of students at the University of California, Davis, for over a decade. This reader was continuously revised over the years by one of the authors (PBM), with the help of others who co-taught courses with him, in particular Douglas Kelt. The book thus rests firmly on the shoulders of many fine students (now practicing conservation scientists), who contributed directly to the various versions, including Anne Brasher, Jay Davis, Steve Ellsworth, Robert Meese, Mary Orland, Anitra Pawley, and many others. We thank them for their skill and clarity of thought that laid the foundations for this new book.

The authors acknowledge the following individuals, all of whom contributed in some way and without whom this book would never have been made: Sarah Zaner (a most outstanding non-scientist editor, sounding board, cheerleader, and all-around best friend), Seth Riley (a great inspiration as a truly dedicated conservation practitioner), Sudeep Chandra (who field tested and provided comments on early versions of the current book), Peter Hodum (who lent his considerable encouragement throughout the years), Sean Lema, Greg Cunningham, Elena Berg, Wayne Summers, Mark Rains, Karen Holl, Cheryl Brehme, Chris "Sid" Nice, Jackson Shedd, Jason Hokesema, Tag Engstrom, Donald Miller, Daniel Rivers, Katie Rivers, Boh Rivers, and David Marchetti. We also thank our editor at UC Press, Chuck Crumly, for the help and encouragement it took to get the book off the ground. Both authors also thank their respective institutions for encouraging and supporting the considerable work that has gone into making this book a reality. Many academic institutions do not support this type of educational research activity, but thankfully both of ours do, and we are grateful for it. Lastly, we thank the hundreds of non-science students who have come through our classrooms over the years: the enthusiasm, creativity, and energy you have for this subject is the fuel that drives our love of teaching.

Environmental History

Human interactions with the environment are constantly changing. This is well illustrated by the history of human-wildlife interactions in North America, a continent that has been inhabited by us humans for "only" the past 13,500 years. Our history here is also a brief history of the concepts of *nature, wilderness,* and *wildlife.* A discussion of the changes in meaning of these seemingly simple terms highlights the dramatic changes that have occurred in our understanding of the world we live in. The history of these ideas reflects our changing attitudes toward the environment in a broader sense and helps to illuminate how we have gotten ourselves into the present environmental crisis. An understanding of environmental history also provides reasons to think that there is at least some hope we can work our way out of the present crisis with many of our natural systems intact. In order to better delineate such a long period of history, we break the historical record up into somewhat arbitrary "eras" that are borrowed from *An Introduction to Wildlife Management* by J. H. Shaw (1985). The exact dates for these eras have little meaning by themselves but

instead act as indicators of cultural shifts that are largely continuous in nature. This chapter serves as an introduction to the entire book. We introduce ideas, terms and topics that we will revisit many times in the coming chapters. Our goal is to provide a basic understanding of the historic roots of modern conservation biology and of the global environmental crisis.

PRE-EUROPEAN ERA (11,500 BC TO AD 1500)

Humans invaded North America some time during the last ice age, roughly 13,500 years ago, when sea levels were lower and it was presumably possible to walk across the Bering Land Bridge connecting Asia and Alaska. Although evidence is scanty, it appears that once the glaciers melted sufficiently to allow passage out of Alaska, colonizing bands of human beings spread across the continent. Using boats, they apparently moved along the coast, down to the tip of South America, in less than 1,000 years. Even at this early stage of the human invasion, there is evidence to suggest that people had a major impact on their environment and the wildlife with which they lived.

Before humans entered the picture, North America had an impressive assortment of very large mammals and birds. Many members of this group were plant eaters (herbivores) and included elephants (woolly mammoths, giant mammoths, and mastodons), horses, camels, giant bison, giant ground sloths, giant armadillos, tapirs, giant beavers, giant tortoises (roughly the size of a small car), and a giant pig that was as large as the largest boars of Europe. An entire group of now extinct megapredators preyed on these large herbivores, including cheetahs, saber-toothed tigers, giant wolves, and two species of lion (one larger than the modern lions of Africa). There also existed a truly fearsome short-nosed bear, about twice the size of a modern grizzly bear, which ran down its prey as cheetahs and wolves do today. Jaguars lived far north of their current tropical latitudes, into the forests of Canada, as did many of the New World cats now restricted to Central and South America. There also existed a group of large meat-eating birds, the largest of which were the teratorns, scavenging birds with wingspans up to 5 meters. The endangered California condor is the last remnant of these giant scavenger birds. There was even a giant vampire bat adapted to feeding off the blood of these enormous beasts. So, why isn't North America still home to this wonderful array of creatures?

PLEISTOCENE OVERKILL

The fate of the giants of the past has been the topic of much debate, but considerable evidence supports a hypothesis called "Pleistocene overkill." The idea is that, as humans spread across North and South America, they preyed upon the large herbivores, such as mammoths, ground sloths, and tortoises, and wiped them out. As originally formulated by Paul Martin, the idea was that the large mammals were driven to extinction in a few hundred years in a blitzkrieg-like event. A newer version of the hypothesis is that the extinctions were more gradual, based on evidence that, in some areas, humans and large animals coexisted for long periods of time, despite hunting. However, the end result was the same: extinction of the megafauna. Large animals are more vulnerable to extinction than smaller ones because they cannot hide easily from human predators and because they reproduce quite slowly. It is possible that the large animals were also relatively unafraid of human beings, because they would have evolved for hundreds of thousands of years without humans present. In addition, there is some indication that a rapid shift in climate reduced the habitats of many of the giant herbivores, making them more vulnerable to human predation. Likewise, Australian biologist Tim Flannery suggests humans may have changed the environment through their actions, especially by increasing the frequency of fires.

Not unexpectedly, when the large herbivores disappeared, their natural predators, such as saber-toothed tigers and short-nosed bears, became extinct as well. The large scavenger birds, which had adapted to eating the remains of large animals, also followed them into extinction. The California condor may have held on because it had access to the carcasses of large marine mammals such as whales and sea lions, which did not go extinct at this time. The loss of these giant animals also impacted the diversity of smaller animals. Because abundant large animals (such as mammoths and tapirs) alter plant communities by the way they graze, their disappearance would have caused a shift in the plant communities, resulting in the extinction of many smaller species that depended on the habitats maintained by the large grazers. In fact, there existed a grassland ecosystem in Alaska called the *mammoth steppe* that disappeared entirely once the woolly mammoth went extinct in that region.

The idea of the Pleistocene overkill is quite controversial, yet the principal alternative to explain the rapid loss of all these giant animals is a drastic climate change that occurred with the end of the ice age. Recent fossil data and archeological discoveries increasingly support the idea that the first native peoples were responsible for the extinction of many species.

One of the early groups (but not the first) to colonize North America was the Clovis people. At Clovis archeological sites, researchers have found distinctive, beautifully made stone spearheads that would have been well suited for killing large herbivores. These same Clovis spearheads have been found imbedded in the skeletons of large animals, which strongly suggests that these animals were hunted by the Clovis people. Clovis culture rapidly spread throughout North America, and then rather abruptly disappeared after about 300 years—a disappearance that seems to coincide with the extinction of many (but not all) of the large animals of interior North America.

Additional support for the overkill idea comes from data from the growth rings of mammoth tusks at the time of the mammoths' extinction, which indicate that the animals were eating well and not experiencing the starvation that would normally accompany climate-driven extinction. Also, many of the extinct species of megafauna had already survived several other glacial climate cycles during the previous 100,000 years, and so presumably they could have survived one more. It is worth noting that a similar event occurred in Australia, with early humans apparently wiping out a suite of giant marsupials, as well on islands throughout the world. If we accept some version of the Pleistocene overkill hypothesis, then we also have to accept the idea that early hunter-gatherer societies were capable of having large and permanent effects on their environment.

DISEASE AND WILDLIFE POPULATIONS

When the first Europeans arrived in North America they were often impressed with the abundance of wildlife. For example, when Daniel Boone brought colonists to settle the Ohio Valley, he led them into a wilderness of large trees, teeming with deer and bear. Two hundred years earlier, however, this same valley had been largely cleared for farms, which were tended by a dense population of native people. The main cause of the disappearance of so many of the First People at the time of European contact was disease.

Disease epidemics are most likely to occur in dense populations of animals or humans because pathogens can spread more easily among individuals and there is a large supply of susceptible hosts for the disease. Not surprisingly, high population densities in urban centers of Europe and Asia supported many nasty communicable diseases, such as smallpox, measles, chicken pox, bubonic plague, malaria, cholera, yellow fever, and influenza. Most Europeans were relatively immune to diseases such measles and smallpox because they had essentially evolved with the diseases, many of which originated in pigs and other livestock. The plagues such as Black Death that periodically swept through Eurasia, killing millions of people, were powerful agents of natural selection (see Chapter 2). When Europeans came to America, they brought these diseases with them; unfortunately, the native peoples had virtually no resistance to the diseases. Smallpox, measles, and other diseases carried by the early explorers, from Columbus onward, rapidly swept through the continent, dramatically reducing Native American populations. Cortez's conquest of Mexico, for example, was greatly hastened by the reduction of the Aztec population by a measles epidemic.

To say that the effect of these illnesses on the population of the Americas was devastating would be an understatement. It has been estimated that, by the end of the seventeenth century, between 70 and 90 percent of First People had died from European-imported diseases. When the native peoples were wiped out, their game species responded by increasing in number. If you remove major predator, such as humans, then prey numbers will increase, at least temporarily. Thus, the first impact of Europeans on wildlife in the Americas was probably to increase wildlife populations through the tremendous and tragic reduction of the populations of indigenous peoples.

BELIEFS AND ATTITUDES TOWARD NATURE AND WILDLIFE

Many of the earliest North American peoples were hunter-gatherers, meaning that they collected most of their food and were often migratory or nomadic in their behavior. Other groups were more sedentary but managed the local landscapes to produce the goods and services they needed, using tools ranging from fire (to favor desirable grasses and shrubs), to irrigation of crops, to regulating the catch of salmon coming up rivers. No matter where the native peoples were on the spectrum of environmental manipulation, they were dependent on understanding the natural world for their survival. This direct dependence on natural ecosystems required an intimate knowledge of the natural world that was often reflected in their beliefs and attitudes toward nature and wildlife. Such peoples commonly viewed themselves as inseparable from the natural ecosystems and wildlife around them. Animals were often regarded as another "kind" of people, or as spirit beings, who could be appealed to for help and protection. Rituals were commonly performed to show respect, gratitude, and reverence for the animal spirits, with the hope of promoting continued harvest success. Other rituals served to influence natural events, such as the coming of rain, or the harvesting of a crop such as acorns, maize, or wild grains.

The belief in magic, ritual, and the intimate fusion of humans with the natural world can be seen in the cultural traditions of many native cultures, as well as in the wonderful portraits of bison, deer, bighorn sheep, salmon, and other animals painted on cliff walls or in caves. As a species, we *Homo sapiens* have been hunter-gatherers for most of our evolutionary history; in this time, we did not separate nature, wilderness, and wild animals from human culture. Some anthropologists suggest that, as a result of this connection to nature and wilderness, we

experience deep-rooted psychological feelings of well-being when we are placed in circumstances that mimic our ancestral environments. Such environments include natural settings (such as forests and prairies) with views of wildlife (or domestic animals), often near water, especially when frequent interactions with close family members are involved.

ERA OF ABUNDANCE (1500–1849)

Early European descriptions of the New World emphasized the incredible abundance of plant and animal life. The Europeans were astounded by the sheer number and variety of fish available. They were left speechless by the flights of passenger pigeons and measured their numbers in the "millions and millions." Accustomed to firewood shortages in England, settlers were delighted by the variety and abundance of trees in the forests. Unfortunately, in the settlers' minds, the new landscape also contained native peoples, even if their populations were in low numbers because of disease epidemics.

When European settlers came into contact with the indigenous peoples of North America, four factors, all of them related to perceptions of animals and nature, allowed them to justify taking from the natives. First, the unsettled areas of the continent were regarded as inhospitable wild areas that were feared and that needed to be tamed to allow them to be used. Thus, conversion of wilderness to farmland and cities was regarded as a good thing.

Second, settlers generally thought that the native people who lived on the land were not using the land properly. They did not "own" the land in the European sense. Because the native urban cultures had largely disappeared, the native peoples the Europeans encountered were largely hunter-gatherers or practitioners of shifting agriculture, meaning that they would clear a field and then move on when the field began to lose its fertility. European farmers, on the other hand, were accustomed to a system in which a person would farm the same piece of land "in perpetuity," regardless of the health or fertility of the soil. Europeans viewed Native Americans, as "lazy," because they engaged "only" in hunting and fishing. Such activities, in Europe, were leisure activities, not integral parts of a cycle of the seasons as they were for most of the First People. It was much easier to justify taking land from people who were lazy, who didn't own the land, and who were not living up to the biblical commandment to subdue the Earth.

Third, the devastation of native peoples by European diseases (when it was recognized at all) was regarded as an act of Providence, favoring the obviously superior European culture. In the wake of depopulation and social chaos, taking "unoccupied" land seemed not only justifiable but prudent.

Finally, Europeans perceived the environment of the New World not as a set of integrated ecosystems but as a collection of commodities. These commodities, in incredible abundance, were apparently free for the taking. In addition, they saw not only the land but the animals and plants that lived on it as private property. Whereas in most hunting and gathering societies an animal belongs to the individual who kills it, in European society, wildlife was owned by the people who owned the land. When Native Americans "sold" land to Europeans, they perceived themselves as merely sharing the use of the land. Europeans perceived themselves as buying the land and everything on it, regardless of the way the land might be used. These ideological and economic differences caused conflict repeatedly and still do to a certain extent. This can be seen in the history of beaver trapping.

THE BEAVER TRADE

The beaver trade was stimulated by the need of the early European colonies to find a commodity that would repay the debts they owed to European merchants. European settlers and traders were quite aware that they were not as efficient as native hunters, so they hired native people to do the hunting for them. As "lazy" as native peoples were perceived to be, Europeans admitted that they were superb hunters and trappers.

Traditionally, hunting peoples had traded with agricultural peoples on the southern coast of New England, exchanging maize (corn) for pelts. Europeans inserted themselves into the traditional network, initially using *wampum* (shell beads) as currency. However, the context was very different: traditional Native American trade had been based on a complex network of kinship and friendship and had primarily been local, from village to village (although some peoples also made long trips for trading purposes, to acquire precious goods such as obsidian or copper). The new trans-Atlantic fur trade stretched over long distances. However, as a result of native populations being dramatically reduced by disease, traditional kinship groups had largely broken down, and the original networks of trade were weakened.

The idea of beaver pelts as a commodity, as something that could be removed from the environment with no consequences, had massive ecological and social repercussions in New England and throughout the beaver range. Earlier, Native Americans had had little incentive to kill more animals than they needed. They never accumulated animal skins beyond the need for personal use and a little barter. In many tribes, all of a person's possessions had to be moved many times a year as the village or family followed animals seasonally. Because the native people lacked large draft animals, except dogs, everything a family owned had to be carried on family members' backs. Pre-European trade thus was inherently conservation oriented. When the traditional trade practices were altered by European colonists, the disintegration of earlier, more ecologically friendly practices followed.

In New England, it was clear by 1640 that beaver numbers had declined. By 1650 the trade in Massachusetts was described by its founder, William Pynchon, as "of little worth." Nevertheless, from 1652 to 1658, Pynchon's son managed to procure 9,000 beaver pelts as well as hundreds of moose, otter, muskrat, fox, raccoon, mink, marten, and lynx skins. By the end of the seventeenth century, the beaver trade was dead in New England, yet the fur trade continued through the eighteenth century. When one area became trapped out, hunters and trappers would move farther inland, especially into the difficult terrain of interior Canada. By the end of the eighteenth century, the fur trade became no longer profitable, in large part because beavers and other fur-bearing animals had become extremely scarce across North America.

The conservation implications of the beaver trade went far beyond the extinction or decrease in the range of the species. As beavers disappeared, eventually the ponds held behind beaver dams became full of silt, and the dams collapsed. The rich soil that was exposed was prized by settlers for its agricultural and pasture potential. The destruction of beaver populations, in much the same way as the epidemics, actually opened up land for European settlers. Westward moving beaver trappers were followed by farmers, who sought out the good meadowlands around streams that had once supported beavers.

Another consequence of the near extinction of the beaver was major changes in the plant and animals communities in and along streams. Beavers cut down trees, reducing shade and increasing the diversity of smaller plants, which in turn provides food for rodents and other herbivores. Naturally abandoned beaver ponds become meadows, which support elk, deer, bear, and other large animals. The beavers' actions increase the complexity of both the terrestrial and aquatic environments around them. Thus, the disappearance of the beaver from an area may result in the decline of other large animals, as well as their predators. In the western United States, for example, it is likely that declines of coho salmon and other migratory fish ware hastened by the shortage of rearing habitat for juveniles that had once been provided by beaver ponds.

By the time the beaver trade collapsed, many native communities had changed beyond recognition. Instead of producing most of the goods necessary for survival, the people hunted and trapped fur-bearing animals and sold all of the pelts they acquired. They became dependent upon European trade goods such as blankets, fabrics, and food. By the 1660s, with the beaver gone, the native peoples of New England turned to the one commodity they had left to sell, their land. Those few who had survived epidemics, loss of income and trade, and loss of land began to keep European livestock. In the far north where the beaver trade continued, native peoples began to accept European notions of animals as property. Territories used by particular bands became more fixed in an attempt to conserve and ration the beavers that were left.

It is ironic that today the combination of decline in the value of beaver pelts and the protection of beavers has resulted in an explosion in beaver populations, including some in urban areas. Beavers are often regarded as pests because they block drain culverts with dams, which results in the flooding of roads and parks, because they burrow into levees and weaken them for flood control, and because they cut down trees planted for home landscaping or stream restoration projects.

The decline (and resurgence) of beaver populations demonstrates the unanticipated consequences of intense exploitation of a key species to the natural landscape for a host of other species, including humans. Their loss is symbolic of the profound changes that took place in North America as the landscape switched from being dominated by indigenous peoples to being dominated by Europeans. It was also an early example of changes that became commonplace during the "Era of Overexploitation."

ERA OF OVEREXPLOITATION (1850–1899)

This era was one in which the North American continent was transformed from a land of apparently limitless wilderness and natural abundance to one with cities and farms scattered everywhere, held together by a spidery network of railroads, roads, and telegraph wires. This time period saw rapid settlement of the West Coast (catalyzed by the discovery of gold in California); the Civil War; the virtual disappearance of eastern forests; an enormous influx of immigrants from Europe, Asia, and Africa;

FIGURE 1.1. These three woodcuts reflect a nineteenth century view of nature as something that needed to be controlled or tamed, exaggerating its fearful aspects with imagery such as an eagle carrying off a small girl (no eagle is large enough to do that!). (These are a few of many similar illustrations from a popular book on natural history by Buel, J. W. 1889. The living world: a complete natural history of the world's creatures. Holloway Publishing. St Louis, MO.)

and a vast expansion of industry and technology. The large increase in human population during this era, combined with the technology of early industry and the demands of a market economy, caused wildlife populations to plummet from a combination of overexploitation and environmental alteration. Some of the many examples include the following:

- The vast migratory herds of bison on the Great Plains were systematically slaughtered or died of cattle-borne diseases until only a few hundred individuals were left.

- The passenger pigeon, whose numbers were once reckoned in the billions, became extinct in the wild. Both adults and young were harvested commercially. The last bird died in captivity in 1914.

- Heron and egret populations were razed by hunters shooting them in their breeding colonies for plumes for ladies hats.

- The ranges of large predators such as grizzly bears, mountain lions, and wolves became greatly reduced. Mountain lions and wolves were virtually eliminated from eastern North America, as were grizzly bears from California (found on the state flag no less) and most of the rest of the western United States.

- White-tailed deer became extremely scarce in the eastern United States through a combination of habitat loss and overhunting.

- Runs of salmon and shad disappeared from most eastern rivers; their runs were blocked by mill dams, or they were killed by factory wastes in combination with unlimited fishing.

The drastic decline of wildlife is not really surprising, considering the attitudes of most people living in this era. Nature was regarded as something that got in the way of civilization and progress, and as a source of goods to sell on the market. People were often frightened by the abundant wild animals and uncontrolled wild ecosystems and thought nature and wilderness had to be tamed and controlled. Thus, popular nature books of the era were filled with drawings of animals doing nasty things to people or to each other: bears clawing hunters, eagles carrying off children, deer goring one another, land crabs attacking goats (Figure 1.1). One of the worst examples of overexploitation during this time is illustrated by the story of the American bison.

BISON AND POLITICS

Bison, often referred to (incorrectly) as buffalo, are one of the most enduring symbols of American wildlife and the "Wild West." Enormous herds of bison on the Great Plains could number in the millions, and it has been estimated that the total bison population in North America at one time reached over 30 million individuals.

By the end of the seventeenth century, the new United States had ceased to be centered in the colonies of the Atlantic Coast, and the westward expansion of European settlers had begun. Settlers often held a belief that they were divinely appointed users of the American earth and that this use was for the good of all humankind. Politicians of the eighteenth and nineteenth centuries argued that the native peoples needed to give up their territory because they had no use for it except hunting, gathering, and fishing and they should adopt the "superior" European cultural ways. Even agricultural peoples who held land in common, such as the Cherokees of the southeast, were not considered to be proper farmers. In fact, native people were often compared to animals; they were believed to be genetically inferior to Europeans; their cultures were believed to be "savage." The word "wild" was often used to refer to native people, especially to those who refused to give up their way of life to partake of that of Europeans. This came to be the dominant language among both American political thinkers and people who were taking part in the westward expansion. As a result, any action was justified, as long as it had the effect of removing "savages" from land that obviously would be better used by Europeans. This included removing natural food supplies. The most blatant example of this process was the destruction of the bison herds by white hunters.

The descriptions of bison by early European explorers are reminiscent of the words of settlers who first came to New England and described the native wildlife. Millions of the creatures inhabited the area between the Mississippi River and the Rocky Mountains. George Catlin, the famous nineteenth century American painter, commented in his journals that the herds he saw stretched as far as the horizon.

The native peoples of the Great Plains began hunting bison intensively in the early eighteenth century, almost as soon as they had acquired horses. A number of tribes gave up their earlier, more settled existence to follow the herds. By the time of European settlement 100 years later, their culture was almost entirely dependent upon the bison. Though older religious and social institutions still existed, it is clear from the customs, songs, and kinship networks that bison were the center of these tribes' lives. Everything from tents and shoes to glue was either derived from the bison or from smaller animals that were hunted in addition to it. Virtually every part of the animal was used, unless large numbers were killed at once.

In the early history of European settlement, the East Coast colonists had no real interest in settling in the Great Plains. Explorers such as Lewis and Clark and those few settlers who crossed the Plains on their way to Oregon, California, or the Southwest generally described the Plains as a "howling wilderness," best traversed as quickly as possible. In the 1830s and 1840s, few Americans believed that the native people who lived on the plains were much of a problem. The critical events that changed this attitude were the discovery of gold in California in 1849 (and subsequent discovery of silver in Nevada and gold in Colorado) and the completion of the transcontinental railroad 20 years later. In order to exploit the new mineral commodities, it was necessary to get through the Plains and past the Native American "savages," who roamed across the area.

In the late 1800s, the U.S. government used three tactics to clear the First People out of the way: (1) military attacks, which were only intermittently successful, (2) deliberate spread of epidemic disease to isolated groups of peoples (through infected clothes and blankets), and (3) destruction of the bison herds. Particularly in the two decades following the Civil War, hunters literally killed bison by the millions. Sometimes carcasses were left to rot; sometimes the animals' tongues and hides were taken. Two to four million were killed each year during the 1870s. Twenty thousand hides were sold in St. Louis in one day. Even the animals in Yellowstone did not escape the slaughter. During the 1870s and early 1880s, bison were killed there by the thousands. Like other game animals, bison were killed and eaten by park employees. In 1882, to save money on beef, one of the first concessionaires in Yellowstone hired professional hunters

to provide 20,000 pounds of elk, deer, mountain sheep, and bison meat. Only the intervention of the U.S. Cavalry in 1886 and its effort to stamp out poaching in the park saved the last 25 bison and other large herbivores from being exterminated from our first national park.

By the early 1880s, native peoples could no longer find bison in the numbers required to sustain themselves. Like the people of the East Coast, with their livelihood gone, they became increasingly dependent upon imported American and European goods and foods. A few escaped to Canada, where they continued to follow the diminishing bison herds, but, in desperation, many agreed to sign treaties ceding the majority of their land to the invaders in return for food and clothing. By 1889, 85 wild bison were left outside Yellowstone National Park, and the last of these was shot in Colorado in 1897. By 1902 there were only 20 individual bison left in Yellowstone, and 150 in Canada. The species had gone from over 30 million to less than 200 in a few decades under the onslaught of early western industrial society.

Today, there is growing recognition of the devastating impact the loss of bison had not only on the native peoples but on the Great Plains ecosystem in general. As the bison herds slowly rebuild on both public and private lands, so does the appreciation of their adaptation to the prairie environment and of the potential to recreate large open areas of grazing land from which bison can be harvested in a sustainable fashion. The creation of a "buffalo commons" has the potential to provide enormous benefits to the land, the prairie ecosystem, and the people who live there.

ROOTS OF THE CONSERVATION MOVEMENT

Despite this dismal picture, the Era of Overexploitation also contained roots of the modern conservation movement. Over the clamor of self-congratulation for having subdued the remote, barren, rocky, bushy, wild-woody wilderness into a second England, a few individuals asked if this transformed environment was what people really wanted. Henry David Thoreau in 1855 sat down with his journal beside Walden Pond, the local swimming hole in his hometown of Concord, Massachusetts, to consider the ways in which Concord had been altered by two centuries of European settlement and expansion. Taking a 1633 account of the area and comparing it to what he saw around him, he concluded that the changes had been drastic. In *Walden* he listed animals and trees that were no longer present in Concord in 1855, and then wrote:

> I take infinite pains to know all the phenomena of spring, for instance thinking that I have here the entire poem, and then, to my chagrin, I hear that it is but an imperfect copy that I possess and have read, that my ancestors have torn out many of the first leaves and grandest passages, and mutilated it in many places. I should not like to think that some demigod had come before me and picked out some of the best stars. I wish to know an entire heaven and an entire earth.

Thoreau was not the first to comment on the changes in the environment he saw around him. Unfortunately, however, until the twentieth century, writers like Thoreau could find almost no audience. Most Americans at the time believed strongly in the idea of Manifest Destiny, whereby the United States was seen as having a divine imperative to conquer and subdue the lands from coast to coast and therefore had little interest in writers like Thoreau. People homesteading on the plains, if they read Thoreau at all, would have thought of him as one of the educated elite who knew nothing of their struggles to turn wilderness into productive farmland. It was only later that Thoreau's genius was recognized and his philosophical stance held in high regard.

Also during this era, Charles Darwin published *On the Origin of Species* in 1859, which further fueled the arguments that humans were part of nature, not separate from it, and that humans had a great impact on the natural world, including causing the extinction of species. Such ideas were not quickly absorbed by mainstream America, but people in this era were slowly becoming aware that the country was losing its natural heritage. Thus, during this time, the first game wardens were hired, some states began requiring hunting licenses, fish and game commissions were established to find ways to improve hunting and fishing, and Yellowstone National Park was established.

Yet even these efforts were based on a philosophy that humans could improve upon nature and that wilderness and wild things needed to be under human control. Thus, the newly established fish and game commissions often took their major task to be the introduction of new species. The largest railroad cars that existed in this era were those designed to carry fish back and forth across the continent so they could be stocked in places that needed improvement. Striped bass and American shad were introduced to California from the East, with the cars bringing back rainbow trout and Pacific salmon for introduction into eastern streams. Carp from Europe were introduced everywhere and were considered to be so much better than native fishes that, for a while, pools beneath the Washington Monument in Washington D.C. were used to rear them.

ERA OF PROTECTION (1900–1929)

At the turn of the twentieth century, most Americans were still rather oblivious to the environmental deterioration that was occurring everywhere, but some of them were outraged by the uncontrolled hunting that was eliminating populations of the most spectacular animals, from deer to herons. The focus of this era was to gain some minimal level of protection for desirable animals, in part so they could continue to be harvested.

The era began with two significant events: the passage of the Lacey Act in 1900 and the accession of Theodore Roosevelt to the presidency of the United States in 1901 (Figure 1.2). The Lacey Act helped to eliminate market hunting for plumes from birds (by prohibiting interstate commerce in feathers) and was passed

FIGURE 1.2. Teddy Roosevelt and John Muir at Glacier Point, above Yosemite Valley, California, in 1903. As president, Roosevelt protected vast stretches of the American wilderness, including parts of what would become Yosemite National Park. John Muir's writings and activism helped form the Sierra Club and protect Yosemite Valley and Sequoia National Park. (Library of Congress Photo.)

in part due to lobbying from newly formed Audubon Societies. The period of the Theodore Roosevelt presidency is considered by many to be a golden age of conservation. Roosevelt, an ardent hunter and conservationist, tripled the size of forest reserves to 148 million acres and created the U.S. Forest Service to manage and protect them. He pushed Congress into passing legislation that withdrew 80 million acres of public land from exploitation for coal and 4 million acres from exploitation for oil. He created the first national wildlife refuge (which snowballed into the National Wildlife Refuge System), created many national parks and monuments, and beefed up federal enforcement of wildlife laws. Roosevelt also used the 1906 Lacey Act to set aside several million acres of public land as national monuments. In 1913 and 1916, laws were passed that essentially made

hunting of most migratory birds, except water-fowl, illegal.

The conservation movement in the United States continued to grow in the early part of the twentieth century. Some individuals such as John Muir (Figure 1.2) realized that mere preservation of wilderness areas was not enough. He helped found the Sierra Club in 1892 in an attempt to get ordinary people involved in and educated about wilderness. Why, Muir asked in his writings, was there such a dichotomy between lifeless cities and untrammeled wilderness? Could city people care about wilderness they might never see? Muir thought they could. The Sierra Club advocated the establishment of Yosemite National Park in John Muir's beloved California Sierra Nevada mountains—what he called the "Range of Light." The Sierra Club attempted to preserve not only the Yosemite

FIGURE 1.3. Artist's re-creation of Hetch Hetchy Valley, Yosemite National Park, California, before it was filled with water by O'Shaugnessy Dam in 1923 to provide water and power to San Francisco. Currently this lush valley is filled with thousands of gallons of water. (Used with permission of Restore Hetch Hetchy.)

Valley itself but also the high country surrounding the valley and the neighboring (and equally beautiful) Hetch Hetchy Valley (Figure 1.3). Although Muir and the Sierra Club succeeded in protecting the former, they lost the latter. Hetch Hetchy is now a reservoir owned by the City of San Francisco that supplies power and pure mountain water to the city.

The flooding of Hetch Hetchy Valley was symbolic of the basic attitude of resource managers and the general populace in this era: nature could be improved upon in order to yield its products to humans in greater abundance. Thus, introductions of species continued unabated, and state and federal governments initiated major programs in predator control. Predators such as mountain lions, bears, wolves, coyotes, and foxes were considered to be vermin that should be shot, poisoned, and trapped in order to increase populations of game animals such as deer and elk and to reduce predation on livestock. In fact, taxpayer-funded predator control programs were still in effect at some state and federal management agencies until the Clinton administration of the late 1990s.

During the Era of Protection, there was an increasing consciousness of the value of natural ecosystems and wildlife for their own sake, tempering the idea that humans could improve upon nature. However, ecology as a science was only in its beginnings, and many of the haphazard attempts to improve upon nature by resource

managers were beginning to backfire. The next era would be marked by improvement in scientific understanding of wildlife populations and ecosystems and by a continuing increase in the valuing of wildlife.

ERA OF GAME MANAGEMENT (1930–1965)

This era began in 1930, when the Report of the Committee on North American Game Policy was issued. The committee, chaired by Aldo Leopold, made strong recommendations for better research on and management of game animals. Leopold was the founder of the University of Wisconsin's Department of Wildlife Management, and in 1933, he published *Game Management*, a book which is often used as the milestone heralding the birth of wildlife biology as one of the precursors of modern conservation biology. Throughout this era, there was a growing awareness that creatures other than game animals also needed protection. National parks became more regulated with regard to allowable uses and the Wilderness Act of 1964 allowed the creation of wilderness areas on U.S. Forest Service and Bureau of Land Management lands. The growing awareness found its philosophical justification in Leopold's *Sand County Almanac* (1949), which, in elegant prose, outlined the need for environmental ethics and the maintenance of intact ecosystems.

The focus in this era was on improving wildlife and fish populations to satisfy the increasing demand for recreational hunting and fishing. State and federal agencies dealing with wildlife and fisheries were strengthened, and new sources of funding such as duck stamps were found. Excise taxes on guns and ammunition (Pittman-Robertson Act of 1937) and on fishing tackle and boats (Dingell-Johnson Act of 1950) provided reliable sources of funds for research and management of wildlife and fisheries, respectively. Although there was a great deal of money spent on habitat management and restoration, such as for the acquisition of wetlands to be used for waterfowl refuges, a prevailing point of view was that much of the recreational demand could be satisfied by raising fish and game under artificial conditions. The animals so produced were then released into areas where hunting and fishing pressure was intense and wild populations depleted. Thus, many states financed large game farms to produce pheasants, ducks, and quail for hunters. Even more extensive were the fish hatcheries, especially those producing trout and salmon. These were often created in exchange for fisheries lost when dams cut off access to upstream spawning areas or flooded streams. It was optimistically assumed that humans could produce more and better fish in hatcheries than the natural environment could produce.

In 1935 Aldo Leopold and Bob Marshall worked together to found the Wilderness Society. The Society was created to be an activist group that was committed to education, action, and advocacy in favor of wilderness. Bob Marshall committed his own money and time to assess the environmental problems caused by the federal road-building projects of the New Deal. As a result of the Society's work, legislation was passed in the 1960s and 1970s designating roadless areas and wilderness areas, separate from national parks. The Society, along with many other conservation groups, was also involved in lobbying for passage of the Wild and Scenic Rivers Act and the Endangered Species Act.

Aldo Leopold was the philosophical leader of the Wilderness Society. He declared that the Society would promote a new attitude, "an intelligent humility toward man's place in nature." His *Essay from Round River* articulated the idea that all parts of an ecosystem play important roles and that no organism should be removed from an ecosystem. "The first rule of tinkering," he wrote, "is to keep all the parts." In *Sand County Almanac*, Leopold wrote "there are those who can live without wild things and those who cannot . . . . These essays are the delights and the dilemmas of one who cannot." He describes his philosophy as a land ethic, drawing upon the ideas of Thoreau and others: "Perhaps a shift in values can be achieved by reappraising things unnatural, tame, and confined in terms of things natural, wild, and free." Leopold, who grew up as a hunter, saw that shift happen in his own thinking. He described a dying wolf losing the "green fire" from its eyes and later commented, "To be trained as an ecologist is to live alone in a world of wounds."

Between 1940 and 1960, there were few new developments in conservation, in part because the world was strongly affected by the devastation and recovery from the Second World War. As people in the United States and Canada took advantage of postwar prosperity, the exploitation of natural resources accelerated, and urban areas expanded. In 1946, the Bureau of Land Management was formed to administer federal grazing lands and federal lands that had potential for mineral and oil exploration. However, there was much controversy around public lands. The opposition to conservation was led by the timber industry in the Northwest and cattle ranchers in the West. Their views reflected the early industrial view of ecosystems as a collection of commodities. Aldo Leopold's "land ethic" was not only in direct conflict with this earlier mode of thinking: it was also in conflict with the descendants of those early settlers— those who wished to continue to exploit public lands for the commodities (oil, minerals, pasture, timber) they represented. As we shall see, the same conflict is still going on today.

In 1962, Rachel Carson, a biologist, published *Silent Spring* (Figure 1.4). This book jolted the public into seeing that supposedly

FIGURE 1.4. Rachel Carson was a marine biologist and nature writer. Her book *Silent Spring* brought environmental issues to mainstream America in the 1960s, although it was very controversial at the time. *Silent Spring* refers to the widespread killing of song birds by pesticides that was becoming apparent at the time. (USFWS photo.)

beneficial pesticides and other chemicals were producing terrible side effects, most prominently the loss of many species of birds and mammals. The title echoed Aldo Leopold's worried question from *Sand County Almanac* as he watched the decreasing numbers of wild birds: "What if there was no more goose music?" Carson documented the use of pesticides and other chemicals and the pollution of air and water. She showed that the pesticide DDT could not only kill birds directly but could also concentrate through the food chain. "If we keep using pesticides, and if we keep polluting our world," Carson asked, "will we finish the job the first European settlers began? Some day, will there be no more birds singing in the spring?" Her words and Leopold's were prophetic. Research on this question has shown that not only are we exterminating wildlife, but we are also turning the oceans into toxic sumps and may be endangering our own lives by pumping toxic chemicals into the air and into the upper atmosphere. Today, decades after Rachel Carson published her book, her predicted "silent spring" may still come to pass. Migratory birds,

although largely protected from the most toxic pesticides in the United States, are killed by those same pesticides in Central and South America. Both their summer and winter habitats are being destroyed. Millions are killed annually as they smash into wind turbines, fly into brightly lit towers at night, and crash in to the windows of our homes. Meanwhile, subtle new pesticides with subtle new effects are being used.

ERA OF ENVIRONMENTAL MANAGEMENT (1966–1979)

This brief era was a transitional one in which the public, biologists, and other scientists started clamoring for more environmental protection, and politicians reluctantly began to give in to these demands. Its beginning is usually specified as 1966 because the first (but generally toothless) federal Endangered Species Act was passed in that year. This act was strengthened in 1969, and again in 1973, but weakened in 1978. This was the era in which the National Environmental Quality Act (NEQA) was passed (in 1969), requiring environmental impact statements for new projects. In 1970 the Environmental Protection Agency (EPA) was established. Public sentiment was expressed in the extraordinary outpouring of concern seen on the first Earth Day in 1972. Environmental groups grew rapidly. Enrollments in environmental programs at universities skyrocketed, and states began to pass laws similar to the federal NEQA legislation. Even California, with conservative Ronald Reagan as governor, passed a strong California Environmental Policy Act and an endangered species act similar to the federal act.

The prevailing feeling in this era seemed to be that environmental problems could be solved with management that balanced ecological and economic interests. Natural areas could be set aside, for example, to protect species such as kit foxes or kangaroo rats, which were being eliminated by development. The effects of water pollution could be taken care of by reducing discharges or using pesticides that degraded more quickly. The increasingly polluted air of the cities could be cleaned up by people driving slightly smaller cars and by building power plants in the desert.

Between 1965 and 1980, over two dozen pieces of legislation were passed on the federal level, and many times that number on state and local levels, to protect wildlife and wildlife habitats. This legislation included the Safe Drinking Water Act, the Clean Water Act, the Clean Air Act, the Toxic Substances Control Act, and others that sought to decrease pollution. Other pieces of legislation such as the Wild and Scenic Rivers Act and the Surface Mining Control and Reclamation Act attempted to regulate land use and to set aside pieces of land that would be free from development. Wildlife conservation was addressed in the federal Endangered Species Act as well as in similar legislation on the state level. All of this legislation was a direct result of the education and activism in local communities that began to take place in the early 1960s, in large part from the stimulus of Rachel Carson's book *Silent Spring*. Much of this legislation was signed by President Richard M. Nixon.

This was also the era, however, when the United States defoliated huge areas of forest in Vietnam with Agent Orange as part of its military strategy. In the United States, the freeway system expanded, and with it, automobile use and the spread of suburbs across the landscape. Industrial agriculture and timber production, which turned huge tracts of land into crop monocultures, increasingly became dominant land uses. Such actions countered the progress made in environmental management.

ERA OF CONSERVATION SCIENCE (1980–?)

The present era is considered to have begun in 1980 because that was when the first textbook on conservation science (*Conservation Biology: An Evolutionary-Ecological Perspective*, edited by Michael Soulé and Bruce Wilcox) was published. This book gave a major push to the development of conservation biology as a distinct field, a science devoted to finding ways to preserve the diversity of life on Earth. The year 1980 was also marked by the Alaska National Interest Lands Act, which set aside 101 million acres of Alaska as national park, national monument, or national wildlife refuge. This was done because of recognition that the wild lands of Alaska, some of the most pristine areas of the world, were on the verge of being spoiled by mining, settlement, and overexploitation of wildlife. In contrast, 1980 was also the year Ronald Reagan was elected President of the United States; he held a profoundly anti-environmental philosophy. The initial years of this era, therefore, were ones of weakened environmental agencies, confrontational politics on environmental issues, and avoidance of developing serious solutions to major problems such as acid rain.

Despite the attempts to undermine the progress made in solving environmental problems, major improvements have been made. Scientists, and increasingly the public, are realizing that we are experiencing an environmental crisis of global proportions. Human populations are still climbing at an exponential rate, both tropical and temperate rainforests are being cut at alarming rates, and serious pollution is much more prevalent than has been admitted previously. From the perspective of wildlife, this means species are being lost on a daily basis. The biggest advance in this era, however, is the recognition that human activities are having global consequences, most profoundly through climate change and all its ramifications. For example, as a result of the Earth becoming warmer from atmospheric pollution, the ice caps are melting rapidly, and sea level is rising, with enormous potential to flood coastal urban and agricultural areas. A sea change in recognition of the terrifying potential effects of climate change was created in good part by former vice president Al Gore, with his movie *An Inconvenient Truth*. This movie finally made the effects of climate change seem real to most Americans and policy makers.

Recognition of these problems is a precursor for finding solutions to them. The rest of this book will discuss many of these problems and their origins as well as solutions. We have labeled the present era the Era of Conservation Science on the optimistic assumption that our increased awareness of the environmental problems of the world, coupled with our increased knowledge of ecology, will allow us to solve these problems. The questions for you are these: Can the changes in attitude needed to alter our present direction in global use and abuse happen? Is this even desirable? Do the words of Henry Beston, written in 1928, still resonate, or do they represent an old-fashioned attitude, irrelevant in the modern world?

> We need another and a wiser and perhaps a more mystical concept of animals . . . . We patronize them for their incompleteness, for their tragic fate of having taken form so far below ourselves. And therein we err, and greatly err. For the animal shall not be measured by man. In a world older and more complete than ours they move finished and complete, gifted with extensions of the senses we have lost or never attained, living by voices we shall never hear. They are not brethren, they are not underlings; they are other nations, caught with ourselves in the net of life and time, fellow prisoners of the splendor and travail of the earth.

It may be that future generations will label the current period as the Era of Extinction rather than the Era of Conservation Science. The decisions and actions we take today will decide the future.

CONCLUSION

The relationship between humans and nature has changed dramatically over the last 13,500 years, at a pace that accelerates with human population growth. The increasingly large human population requires increasing manipulation of the natural world to make food production efficient enough to feed so many people. The greater efficiency of food production leads to greater social complexity and professional specialization, which in turn leads to even greater rates of cultural and technological change. The increasing need to control nature to provide food, combined with the increasing disconnect between humans and the natural world in daily life, is reflected in views toward nature as our society becomes more and more complex. Thus, during the eras for North America discussed above, wild animals went from being seen as something encountered on a daily basis and appreciated for their natural roles, to being seen as a commodity for sale or a nuisance to be eliminated, to being seen as a precious, irreplaceable resource that must be protected (although arguably much of the human population is indifferent to the fate of wildlife and wild lands). Perhaps the most important trend to keep in mind in the history of the relationship between humans and nature is that the ever-increasing exploitation of natural ecosystems for human use has led to a steady loss of wildlife and wilderness over time, along with the loss of the "free" services they provide to humans, such as fisheries and clean water.

The technological power of modern industrial society makes humans a truly global force, with incredible capacity to manipulate the natural world. The global scale of human power means that we also have a global-scale responsibility to ensure that power is used wisely. There are many aspects of our present society that may foreshadow a sustainable society in the future, such as low birth rates and the increasing high value placed on nature and wildlife. Our society has also brought high standards of living, personal rights, freedom, and economic opportunity to many people, and the globalization of the world's economy may potentially bring these benefits to yet more people.

However, the benefits we enjoy from our industrial economy are in part brought about by rapid exploitation of natural resources, which has significant negative consequences for environmental quality and wildlife populations, and which will only increase in the near future as more countries become industrialized. The potential exists to overexploit the natural resources and systems upon which the human species depends to the point where, in the future, many of the benefits of industrialization will be lost. In the process, many wild areas and wildlife populations are likely to be destroyed, all for no long-term benefit.

The challenge of the twenty-first century will be to figure out how to design a sustainable global society that maintains the benefits of industrialization indefinitely into the future, allows access to those benefits for more people around the world, and still preserves environmental quality and biodiversity. This is a complex and difficult task and will itself be an incredible leap forward in cultural evolution. However, unless we are rapidly able to address these major global issues, it is certain that considerable human suffering and extensive loss of natural ecosystems will occur.

With challenges this large in front of us, it may be easy to feel despair, or to feel that these problems are simply too large for you as an individual to do anything about. However, there are many positive developments that give us hope. Population growth rates have gone to zero or below in post-industrialized nations (with the United States being a major exception) and have already slowed in many developing nations. There are promising new technologies for energy efficiency, non–fossil fuel energy, and carbon sequestration that could greatly reduce the threat of climate change, although bringing these technologies to market requires enlightened and aggressive political support. Standards of living, political freedom, access to education, and human rights have been steadily improving in many countries around the world. Reduced birth rates come with increased education and development. Large national preserves and legislation to protect wildlife and natural ecosystems exist in the United States and in many other counties, although it will take a global effort of managing *all* parts of the Earth to ensure the maintenance of biodiversity. The Internet and the increasingly global nature of human culture aid in the development of international policies to respond to these global environmental challenges. One thing is clear: for better or for worse, the nature of the relationship between humans and the environment long into the future will be largely determined in the next several decades. Whether we create a just, sustainable global society with protected natural areas or a future of climatic catastrophes, wars between overpopulated countries for depleted resources, and near complete loss of wildlife and natural ecosystems will largely depend on the actions taken by the people of our generation.

FURTHER READING

Carson, R. 1962. Silent spring. Houghton Mifflin. Boston. *This classic book still reads well today and the problems she discusses are still with us.*

Diamond, J. 1999. Guns, germs and steel. W. W. Norton and Co. New York. *A highly readable account of how Western civilization profoundly changed the world environment.*

Diamond, J. 2005. Collapse: how societies choose to fail or succeed. Viking Press. New York. *In this book, Diamond deals with historical cases of societal collapse around the globe, many which involve environmental components.*

Flannery, T. 2001. The eternal frontier: an ecological history of North America and its people. Grove Press. New York. *Flannery discusses how natural and human forces have changed the North American landscape through the eons, including the ecological changes wrought by Pleistocene overkill. His 1994 book* The Future Eaters *deals with the same issues in Australia.*

Leopold, A. 1990. Sand County almanac. Ballantine Books. New York. *The all-time North American classic book on why we should conserve the land and its inhabitants. Beautifully written.*

Lott, D. 2002. American bison: a natural history. University of California Press. Berkeley. *An entertaining and personal history of bison in ancient and contemporary North America.*

Martin, P. 2006. Twilight of the mammoths: ice age extinctions and the re-wilding of America. University of California Press. Berkeley. *The evidence for Pleistocene overkill and what it means for conservation.*

Naiman, R. 1988. Animal influences on ecosystem dynamics. Bioscience 38(11):750–762. *An analysis of the historical landscape effects of beaver.*

Shaw, J. H. 1985. An introduction to wildlife management. McGraw Hill. New York. *This book provides an introduction to virtually the entire field of terrestrial wildlife management.*

Snyder, G. 1990. Practice of the wild. North Point Press. New York. *A nice series of essays on the meaning of wild and wilderness by one of America's best poets.*

Warren, L. S. 2003. American environmental history. Blackwell Publishing. Oxford. *An introduction to the historical background of this chapter, featuring excerpts from many classic essays and original documents.*

2

Variation, Natural Selection, and Evolution

Next time you are in a place full of people, stop for a second, glance around, and notice how incredibly varied we human beings are. Hair color, eye color, height, weight, skin tone, shape of the hands, shape of the ears, jaw line, noses, size of feet: wherever you look people are different. Of course, we are really good at recognizing these differences. Now think about how different all dogs look from one another, and that no two horses look exactly alike, or that two trees of the same species growing side by side each have a unique shape and texture. The same goes for house cats, parakeets, goldfish, mountain lions, cows, bullfrogs, rose bushes, tulips, oak trees, carrots, and shiitake mushrooms; for all of these species, each individual animal or plant is distinct. Variation is rampant in nature and exists everywhere we look (Figure 2.1).

Now the question to ask is, why? Why is everything in nature different? Why is there so much variety around us? In this chapter, we will attempt to answer these questions as a way to begin the study of conservation biology. We explore variation in the natural world and the causes of that variation. This will lead you to an understanding of the process of natural selection as well as that of evolution, at both the micro-scale and macro-scale. This understanding is the key to conservation. Protecting species, habitats, and ecosystems requires protecting more than just individuals. It requires allowing variation to continue to be rampant, so that life can continue to adjust to an ever-changing environment.

VARIATION

WHY SO MUCH VARIATION?

So what do you think causes the variation we see around us? Let's take human hair color as an example. What causes the huge variety of hair colors? Have you noticed that, for many people, hair color tends to become lighter when the person has spent extended time in the sun? The environment can therefore affect hair color. What other environmental factors cause variation in hair color? Chemicals (i.e., hair dye) and diet seem to be the most obvious ones. But it is clear that there is some other major force contributing to a person's hair color. In some families, this is very noticeable. You probably know

FIGURE 2.1. Differences in wing patterns in a common butterfly, showing the range of natural variability within species. This variation is the material on which natural selection works. Small differences in wing pattern may make an individual more attractive to the opposite sex or more vulnerable to a predator. (Photo by M. P. Marchetti.)

a family where everyone in the family has the same red (or black or brown or blonde) hair. It's clear that hair color sometimes "runs" in families and is passed down from parents to their children. So genetics also controls variation in hair color.

People tend to lump the causes of variation into two different groups: "nurture" (environmental factors) and "nature" (genetic factors). Environmental factors that cause variation are things such as food, habitat, climate, and lifestyle. Genetic factors include variations in inherited traits ("genes"). But which set of factors is stronger, nature or nurture? Which is more important? Does your genetic makeup mostly determine how you as a person look, or does the environment shape you more? Let's think again about the question of hair color. Which is stronger, genetics or environment? Clearly, the color of your hair is a result of a mixture of both sets of factors. Genes from your parents + hair dye + sun exposure = hair color.

You may be thinking to yourself that, at the basic level, it's really your genes that determine your particular set of variations, but let's think

for a second about variation and inheritance. Is all the variation we see around us inherited? For example, if a friend broke her arm and then had babies afterward, would her kids be born with broken arms? Clearly this is a ridiculous question. The answer, of course, is no. But what if your friend were missing an arm? Would her children be born missing an arm? The answer here is more complex and depends on how or why your friend is missing an arm. If your friend lost the arm in a bizarre gardening accident, then her children would have normal arms. But if your friend was born without an arm due to some genetic abnormality, then there is a chance that her kids could also be born with missing arms. The point here is that not all variation is inherited, but some of it is, and it's that inherited variation we are going to focus on here.

THE LONG AND SHORT OF VARIATION

Let's look at an example in greater detail by exploring the question of whether variation in human height is inherited. To answer this, we gathered data from about 50 students in a biology class by asking them to provide the following information about their biological families: mom's height, dad's height, student's height, and all heights of brothers and sisters. We then compiled and graphed the data (Figure 2.2A) with the *x*-axis being the students' heights and the *y*-axis being the number of students with each height. What do you notice about this graph? First, you should notice that there is a wide spread in the height of the students: 46 centimeters (18 in) separates the shortest and the tallest. Second, a majority of the heights fall around the average height (1.7 m, or 5.6 ft). Finally, there are few students at the extremes (i.e., there are no students 3 or 7 ft tall). This is all well and good, but the question remains: does this tell us whether height is inherited?

Not yet. In order to look at inheritance we need to look at height for both parents and offspring (Figure 2.2B). For each student's family, we took the parents' height and calculated an average of mom's height and dad's height. Then we calculated an average height for all the children in the family. Notice that, for each family, we now have two numbers: an average parent height and an average offspring height. Also notice that these correspond to the *x* and *y* axes on Figure 2.2B. So for each student, we have a dot on Figure 2.2B representing that student's family.

Does anything strike you about Figure 2.2B? Here are some things you should notice. First, there is a trend in the family height data, namely that tall parents tend to have tall offspring and short parents tend to have short offspring. You can see this in the graph because the points all lie in a sort of messy band that increases from left to right. Why are there no points down in the bottom right or the top left part of the graph? What would these points say about parents and offspring? Bottom right points would be tall parents who have very short offspring, and top left points would be short parents who have very tall offspring. Does this graph tell us anything about inheritance of height? It seems that height is indeed inherited (i.e., tall parents

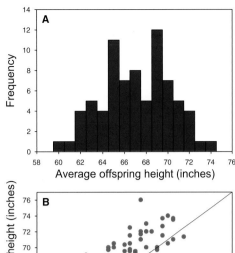

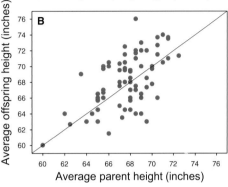

FIGURE 2.2. Graph A shows a distribution of students and each of their heights in inches from an introductory biology class at CSU Chico. Graph B compares the average height (inches) of each family's parents with the average height (inches) of all their offspring. The line in graph B represents the points where average parental height exactly equals average offspring height.

tend to have tall kids), and therefore that height is under some genetic control. But it is not a perfect inheritance. If height were perfectly inherited, then we could exactly predict the height of the offspring by looking at the average height of the parents. We can see this doesn't work by looking again at Figure 2.2B. On the graph, perfect inheritance would be an exact straight line, with all the family points lying on that line. Instead, we see a lot of scattered points, with only a general trend of tall parents having tall offspring.

Why is there so much spread in these data? If height is inherited, why is it not perfect? The answers lie in the environment. Height is indeed an inherited trait, but environmental factors also play a role. One of the most obvious ones is diet. Research has shown that children fed a healthy, high-calorie diet tend to grow taller than children with a calorie-restricted diet. In the United States and the western world in general, people have very high-calorie diets, and as a result, the average height of people in the West has been growing over the last few centuries. There are other potential causes of the scatter in the inheritance graph, including age of the offspring and adoption, but the point here is that *for a simple trait such as human height, the causes of variation are both genetic and environmental in origin.*

WHY AREN'T WE OVERWHELMED BY RABBITS AND COD?

Variation is rampant in the natural world, and some of that variation is inherited. In order to understand why variation is important, it is useful to ask another question: of all the offspring born, which individuals survive? To answer this question, let's turn to the eastern cottontail rabbit *(Sylvilagus floridanus)*. Everyone has heard the phrase "multiplying like rabbits," so we are going to examine what that actually means. For a female cottontail, an average litter might be 10 offspring, typically made up of five males and five females. We want to keep track of the number of rabbits over time, and so to make the calculation easier, we will just keep track of female offspring (remember that females are the only ones who can actually produce more bunnies, so this process does make some sense). If we start out with one female rabbit, and she produces five offspring, then we now have five rabbits (let's assume, for simplicity again, that the ones giving birth die—not realistic, but again, it makes the numbers easier to work with). The average generation time for rabbits is about 16 weeks, meaning that they can breed approximately every four months. So let's keep track of time (generations) and the number of rabbits from the original single female and see what happens after four years.

Incredibly, in only four years, our one female rabbit gave rise to over 240 million rabbits (and remember this is only the females, so including the males would approximately double this number). If this calculation held true, then the

Generation	Number of Rabbits
0	1
1	$5^1 = 5$
2	$5^2 = 25$
3	$5^3 = 125$
:	:
12 (4 years)	$5^{12} = 244{,}140{,}625$ rabbits!!!

entire planet should be completely overrun by rabbits at this point. But, in fact, it is not. The question is, then, what happens to all the rabbits that are born?

To examine this question we turn to everyone's favorite fish for fish and chips—the Atlantic cod *(Gadus morhua)*.

An average female Atlantic cod will spawn approximately two million eggs (in large females, this number may jump to over five million eggs). Given the simple calculation we just did with rabbits and the observation that clearly the oceans are never filled to overflowing with cod, something must be going on. Exactly what happens to all these cod eggs is worth looking into. During the first month after being spawned, 99 percent of these two million eggs die from a combination of causes, including disease, fungal infections, and being eaten. This leaves approximately 20,000 eggs that make it through the grueling first month. Of these 20,000, 90 percent don't make it through the first year of life for similar reasons, leaving only about 2,000 baby cod to reach their first birthday. If we keep track until the baby cod reach the age when they themselves can breed (about three years), we find that, shockingly, an average of only two cod survive out of the original two million, a survival rate of a dismal 0.0001 percent.

The important part of this last sentence may have been overlooked if you were not paying close attention. The key phrase for our purposes is the idea that only two survive *on average.* This means that some females may have six or eight or 100 babies that survive, whereas others may have one or zero that survive. Why is this? Why do some baby cod do better than others? It may be that some come from larger eggs (i.e., variation in egg size) and grow faster, or some may be slightly better swimmers (i.e., variation in swimming performance) and can avoid being eaten. There is also a high degree of luck in who survives; the lucky egg or young is the one not found by a predator. Regardless of the exact reason, it is important to emphasize the radical idea that *most organisms that are born do not survive*. Thus, the question of which individuals survive becomes important.

SNAIL SURVIVORS

In trying to tackle the question of who survives and why, we will turn to yet another gastronomical delight: namely, escargot, otherwise known as the grove snail *(Cepea nemoralis)* (Figure 2.3). These snails are common throughout Europe and were studied by two ecologists from Oxford University, A.J. Cain and P.M. Sheppard. One of the striking things about these snails is that they have a surprisingly variable set of shell colors and patterns. One form of this variation is that some shells are banded (i.e., have stripes) (Figure 2.3), whereas others are non-banded (i.e., have no stripes). It turns out that this form of variation (banded/non-banded) is a heritable trait, much like height is in humans. In general, banded parents give rise to banded offspring. For snails that were just sliming around on the ground, it was found that 47 percent of the population had banded shells, and 53 percent were non-banded. Cain and Sheppard were not studying the snails for shell heritability; instead, they were studying them as a food source for birds. The local birds seem to quite enjoy the snails, eating them by picking them up and dropping them, and thereby cracking their shells and extracting the good gooey bits out of the smashed shells.

Cain and Sheppard started noticing that not all snails are equally likely to be eaten. They watched 863 snails be eaten by birds and noted that 486 of them were banded. This means that 56 percent of the snails eaten by birds were banded, indicating a slight preference for banded snails by birds. What do you think was happening? Is it clear that the birds were preferentially choosing and eating

FIGURE 2.3. The common grove snail *(Cepea nemoralis)*, showing a "banded" shell pattern. (Photo by Atli Arnarson, with permission.)

banded snails, that their choice of snail is not random? If it were random, then they should have been eating 47 percent banded snails because that is the percentage of bandedness in the population. Instead, they were eating 56 percent banded snails. It is possible that the birds eat more banded snails because it is easier to find a banded snail than a non-banded one. Perhaps the non-banded snails blend in to the background and are better camouflaged than banded snails. If we were trying to answer the question of which of the snail offspring survive better, the clear winner here would be the non-banded snails. They do not get eaten as often, and therefore, on average, they will leave more offspring than will the banded snails (note that it is challenging to leave many offspring if you have been eaten by a bird). If this process continues, and all other things are equal, then we would expect that there would be fewer banded snails (because they are constantly getting eaten) and more non-banded snails (because they are leaving more offspring) in the future.

Given that predators prefer banded snails, why do they continue to be just slightly less abundant than non-banded snails? Why don't they disappear from the population? Clearly, the issue is more complex than suggested by the example.

NATURAL SELECTION AND EVOLUTION

This process of differential survival and reproduction we have just described using the garden snail is called *natural selection* and was first described by none other than Charles Darwin. A formal definition of natural selection is *the process by which individuals in a population that are best suited to the environment increase in frequency relative to less well-suited forms, over a number of generations.* Think about the snails. Which ones were best suited to the environment? The non-banded snails. Which ones left the most offspring and thereby increased in frequency? The non-banded snails. What is likely to happen in the future? There will be more non-banded snails around and fewer banded snails. Biologists have noted that there are four general conditions that have to be met for the process of natural selection to occur. They are as follows:

1. More organisms are born than can survive.
2. Organisms within a species vary in their characteristics.
3. Some of the variation is inherited.
4. Because of inherited variation there will be differences in reproduction and survival.

We can see this clearly if we apply this to the snail example:

1. The snails are abundant, but most get eaten before they reproduce.
2. The snails vary in their color pattern (banded vs. non-banded).
3. The variation is inherited (banded snails have banded offspring).
4. There are differences in survival (non-banded snails don't get eaten as much) and reproduction (non-banded snails therefore leave more offspring).

Now that we have a definition and an understanding of natural selection, let's see how this all relates to the idea of *evolution.* At its simplest, evolution can be defined as follows: *a change in the characteristics of a species from generation to generation.* With the understanding that we have developed in this chapter, it is also possible for us to notice that the process of natural selection

can cause evolution. It turns out that there are other forces that can cause evolutionary change (artificial selection/selective breeding, gene flow, genetic mutation, and genetic drift) that we will not dwell on here. Suffice it to say that the process of natural selection can produce evolutionary change.

It is important to note here that Charles Darwin (1809–1882) is one of the most influential thinkers of our age and that his ideas have changed everything from basic biology to medicine to conservation. Darwin did not "discover" evolution, as is commonly supposed. The ideas about evolution and evolutionary change were floating around in the writings of many of the big thinkers during Darwin's time and before. What Darwin did that was so remarkable was to pose the first cogent explanation for how evolution worked. He did this by describing the process of natural selection. This seems pretty straightforward now but it was revolutionary in its scope and simplicity at the time. It was also enormously controversial, so Darwin held off publishing his ideas until he had compiled an immense body of evidence. He not only compiled information about the natural world from many observers and through his own keen observational powers, but he also bred pigeons and other domestic animals to demonstrate how natural selection worked in even artificial environments.

EVOLUTION AND FITNESS

Let's now turn to another useful idea, the concept of evolutionary *fitness.* In the evolutionary context, fitness refers to how well an organism survives and reproduces (i.e., not how long you can work out on a stairmaster). We have all heard the phrase "survival of the fittest," and now we have a context to better understand that phrase. The process of natural selection tends to favor those individuals that are the most fit for their environment. But we have to realize that it is not just the total number of young an individual has that determines an organism's fitness (remember the cod); rather, it's the number of

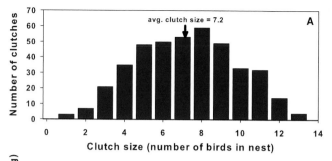

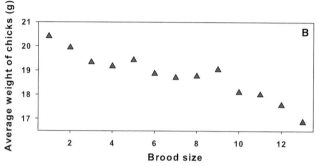

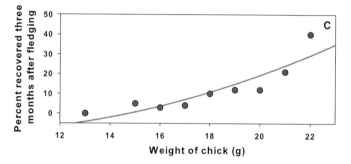

FIGURE 2.4. Distribution (Part A) showing variation in the number of offspring (clutch size) in a population of a small passerine bird called the great tit *(Parus major)*. Part B shows the relationship between the average weight (g) of *P. major* chicks and the number of offspring (brood size). Note that as the number in the brood goes up, the average weight goes down. Part C compares the percentage of chicks that were recovered three months past leaving the nest (fledging) with the chick's average weight (g), indicating that survival and fitness increase with increased chick weight. (Redrawn with permission from Perrins, C. M. 1965. Population fluctuations and clutch-size in the great tit, *Parus major* L. The Journal of Animal Ecology 34(3):601–648.)

offspring that survive and are themselves able to reproduce. Clearly, it is a bit tricky to think about how well the next generation will survive and reproduce, but it should become more understandable with an example.

REPRODUCTIVE OUTPUT AND THE GREAT TIT

We are going to demonstrate how natural selection can work on an organism's reproductive output using a little European bird called the great tit *(Parus major)*. A great body of research has been done on this bird, quantifying many aspects of its life, habits, and reproduction. One of the things measured was the reproductive output of the bird, or the average number of eggs it lays (Figure 2.4A). We can see from the graph (histogram) that the *x* axis shows clutch size (number of eggs in a nest), and the *y* axis shows number of clutches (number of birds that have that particular clutch size). Notice that there is an average clutch size (7.2 chicks) and that the majority of birds have about that many chicks. You should also notice that there is a spread in the data and that birds lay as few as one and as many as 13 eggs (notice that variation is creeping in again). Now we have just said that natural selection favors those individuals that have the highest fitness (i.e., that leave the most offspring), so why are the birds hovering between seven and eight chicks? Why is not 10, 12, or 20 chicks the common value? Is there some limit to the reproductive output of *Parus major*? If we look at a graph of average weight of the chicks against the number of chicks in a brood (i.e., nest) (Figure 2.4B), then we can start to construct an answer. It looks as if there is a rather strong relationship between weight and the number of brood-mates. If there are more chicks in the nest, then the average weight of the birds decreases significantly. This makes some sense when you consider that, in a bird nest, the chicks have to be fed by the two parents, and the parents are only able to get so much food back to the nest every day. Thus, if there are 10 chicks, the food gets divided into 10 parts, but if there are only five chicks, then each chick gets a much bigger share and is therefore likely to grow bigger.

This is still only part of the story. We still do not have enough information to address how this impacts the fitness of the parents. Remember that we said fitness is a measure of the number of offspring that survive and reproduce. So far, all we have done is show that if there are fewer chicks, they grow larger. We still need to tie this to the chick survival in order to get a handle on fitness. Figure 2.4C shows us the relationship between the average weight of the chicks in a nest and the percent of them that were found three months after they flew out of the nest (fledging is the process where birds learn to fly and leave the nest). What do you see here? It again seems that there is a strong relationship, this time between chick weight and survival, where larger chicks survive better. If we tie all three parts of this together, we see that there are good evolutionary reasons for the birds having between seven and eight chicks, on average. There is an evolutionary tradeoff between having more babies and making sure that the babies get enough food to survive when they leave the nest. Seven or eight is the optimal number to balance these competing forces, and this provides *Parus major* with the best strategy to have the highest average fitness.

So why do some great tits continue to produce 11 to 12 eggs and others only five to six? One reason is that the environment is not completely predictable. In some years, insect food may be superabundant, and two parents can actually raise a large number of healthy chicks.

In other years, food may be relatively scarce, so the most successful parents will be those with small broods. Although, in most years, it pays in terms of fitness to be average, there are just enough good and bad years over time to cause variation in brood size to persist.

THE PEPPERED MOTH

Let's look at another example of fitness, this time looking at the genetics of the situation. We will use one of the most common examples in all of biology, namely the peppered moth *(Biston betularia)* in England. Historically, there were two color varieties of the same species of moth, the dark (melanic or pigmented) form and the white form. Note that melanin is a dark pigment that is found both in the moth's body and in our skin. The dark form of the moth was extremely rare (think about how rare albinos are among humans) and accounted for less than 1 percent of all moths (which means that 99 percent of the moths were white). The dark form was so rare because birds had an easy time spotting and then eating the dark moths when they landed on the white trunks of the trees. In reality, the tree trunks themselves were not truly white (in fact the bark is actually black), but they were covered by a white lichen (lichens are plantlike fungi that often cover rocks, trees, and such).

When we say this was the historic situation, we mean that it was the situation prior to the Industrial Revolution, before factories and the use of coal for energy. The smoke and soot from the industrial factories produced a lot of air pollutants, including sulfur dioxide, which forms sulfuric acid when it mixes with water in the air. This acidic airborne brew had the effect of killing lichens on the tree trunks, so the tree trunks were, from then on, black in color. This caused a major shift in the moth population. The white moths, which had been the majority due to their superior camouflage with the lichens, became easy targets for bird predators on the now-dark tree trunks. Conversely, the dark form, which had been rare, became the common form (more than 90 percent of the population).

This is all well and good, but the question we want to ask of these moths is, Does the color variation in the moths get inherited? The answer is yes, and it's a pretty simple genetic system. We need to recall some basic information regarding genetics. Remember that chromosomes are long coiled strands of DNA and that sections of the DNA that code for particular proteins are called *genes*. Also recall that a gene can have alternate forms (think flavors) that are called *alleles*. Now peppered moth genetics are fairly simple, in that for moth color there is one gene with two alleles, a light allele (d—a recessive allele) and a dark allele (D—a dominant allele). Moths are like humans in that we have two copies of all our genes, one copy that came from mom's egg and one copy that came from dad's sperm. Thus, each moth has two copies of the color gene, each with two potential alleles. If we look at making some combinations of the D and d alleles, we can see the following simple pattern in both genotype (what the genes say) and phenotype (the outside or expressed pattern):

Genotype	Phenotype
DD	dark color
Dd	dark color (D is dominant to d)
dd	light color

What we can do with this information is to examine how the percentage of the alleles may have changed over time, considering that we don't actually have genetic samples from that long ago. Remember, we just said that the moth population went from being approximately 99 percent white and 1 percent dark before the Industrial Revolution to being over 90 percent dark and around 10 percent white after the lichens died. The same transformation also happened in the allele frequency of dark (D) and light (d) alleles. The moth population's genes went from being dominated by light alleles (more than 95 percent) before the lichens died to being dominated by dark alleles (more than 90 percent) after the industrial revolution. The point of this example is to highlight the fact that differences in survival and reproduction of the peppered moth led directly to changes in the percentage of alleles over a series of generations. There was *evolution through natural selection* in the moth population. In fact, we now want to slightly revise our simple definition of evolution to include our more sophisticated genetic understanding. *Evolution is a genetic change in the characteristics of a species from generation to generation.* It is a simple alteration, but the definition now more clearly reflects the underlying biology of the process. Evolution involves changes in the genetic makeup of a species. Without changes at the genomic level, we do not have "true" evolution occurring. Evolution means genetic change. Basically, at this point, we have defined the process of *microevolution* (small-scale genetic changes) and will return later in the chapter to explore the process of *macroevolution* (large-scale evolutionary change).

THE CASE OF THE TULE PERCH

A good example of how the fitness of a species is influenced by the local environment can be found in the tule perch *(Hysterocarpus traski)*, a small (4–6-inch-long) fish studied by Donald Baltz at the University of California, Davis. Tule perch occur only in the fresh waters of Central California. Each female tule perch gives birth to 15–40 young, which are essentially miniature adults (they swim away after being born). It turns out that the number and size of young produced by a female is an inherited trait and is an adaptation to the environment in which the perch live. Thus, female tule perch that live in the Russian River produce 25–35 small young and typically become pregnant in their first year of life. In contrast, tule perch in Clear Lake typically produce 15–20 large young and wait until their second year to become pregnant.

The reason for the striking difference in life histories of the two populations is the nature of the environments. The Russian River is a large, isolated coastal stream that fluctuates enormously in flow from year to year. In this harsh system, each adult female has a relatively low probability of survival from year to year, so natural selection has favored females that produce a lot of young quickly (i.e., these are the females with the highest fitness). Clear Lake, in contrast, is a relatively benign environment, where each adult female has a fairly high probability of survival from year to year, provided she is large enough to escape predators. Thus, natural selection favors females that produce large young and that devote all their energy in the first year to becoming as large as possible. If both forms were brought into laboratory aquaria and raised under identical conditions, the Russian River fish would still produce lots of small young and the Clear Lake fish would still produce small numbers of large young.

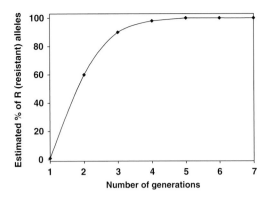

FIGURE 2.5. Estimated percentage of warfarin-resistant alleles (R, dominant) in a hypothetical Norwegian rat *(Rattus norvegicus)* population over a number of generations.

MUTANT RATS

Before we move on to large evolutionary patterns, we want to examine where the variation among individuals and the variation in alleles comes from—to examine the source of all genetic variation. In order to do this, we will introduce another organism, the brown rat *(Rattus norvegicus)*. This rat is a common pest around the world. One of the ways populations of these rats have been controlled is through the use of a rat-specific pesticide (a rodenticide) called warfarin. Warfarin is an anticoagulant poison that is particularly toxic to rats and causes massive internal hemorrhaging and eventual death. Warfarin was developed in 1948 and was first used as a rodenticide in the United States in 1952, where it was initially very successful at killing rats. Rather quickly, however, some rat populations developed a genetic resistance to the rodenticide, meaning that some rats were able to tolerate the poison and not die. By the mid-1970s, resistance to warfarin had spread, and many rat populations were no longer affected by the poison; thus, its use fell out of favor. Interestingly, the genetics of warfarin resistance is very similar to the genetics of the peppered moth example. Resistance is controlled by a single gene that has two alleles, a resistant dominant allele (R) and a non-resistant recessive allele (r). The alleles in the

diploid rats can combine in the same way as in the moths, and we can graphically track the percentage of R alleles in the population. If we had the allele frequency data, they would likely look very similar to Figure 2.5. In this case, there was very strong (i.e., very fast) selection for the resistant allele.

So what happened here? We must realize that the development of resistance is an evolutionary change (i.e., a genetic change in a species over time) through the process of natural selection and that it therefore requires genetic variation in the population. What this means is that, strangely enough, warfarin-resistant alleles were already present in the rat population before the development of warfarin. How is this possible, you might ask? The answer again lies buried in some of that high school biology you had eons ago. Remember that the DNA molecule is amazing in its ability to make copies of itself and that, in this copying process, sometimes small mistakes are made. These mistakes are called *mutations,* and every living being on this planet is carrying around mutations in his or her genomes. Many mutations are bad and may actually kill an organism before it is even born. (Think about a mutation that wouldn't allow cells to use oxygen. How long would an animal like that last?) Some mutations are neutral, meaning that they are neither bad nor good: they are just mistakes in the DNA, sitting there quietly. Some mutations, such as the warfarin-resistance mutation, are neutral until the environment changes (i.e., until warfarin is introduced), and then they give the individual with the mutation a gigantic advantage. Think about it: one lucky rat with the warfarin-resistant DNA out of thousands of rats without it survives the application of the poison and therefore gets to reproduce and pass on its set of lucky genes to the next generation. How great is that? Mutations are the ultimate source of genetic variation and therefore are at the heart of the process of evolution through natural selection.

GALAPAGOS FINCH BEAKS

So far, we have been looking at very simple traits, involving a single gene and only two alleles. The world of genetics is much more complicated than this, and in fact most traits are controlled by many genes and have multiple possible alleles. They are also not of the yes or no variety we have seen in the moths and rats. Human height, for example, is not a two allele system. As we saw, many genetic and environmental factors affect human height. So how does natural selection work on traits like height? To examine this kind of situation, we turn to one of the creatures made famous by Charles Darwin on his world travels, the medium ground finch *(Geospiza fortis)* from the Galapagos Islands. These little birds are seed eaters and have a normal (bell-shaped) distribution of body sizes. The birds with bigger bodies tend to have larger beaks and can crack open larger seeds, whereas the smaller-bodied birds tend to eat smaller seeds. This natural variation in body and beak size is heritable and is therefore passed on to the next generation.

The weather on the Galapagos Islands follows a seasonal pattern every year, with a distinct wet period followed by a distinct dry season. In the late 1970s, a husband and wife team of ecologists (Peter and Rosemary Grant) were studying ground finch populations when the islands were hit with severe drought. There was essentially no wet season in 1977. The finch population plummeted from a high of near 1,200

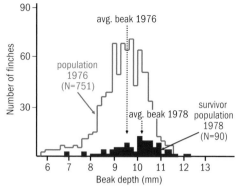

FIGURE 2.6. Distribution of beak size (mm) in the medium ground finch *(Geospiza fortis)* on Isla Daphne Major, Galapagos. Data from the finch population pre-drought (1976) are shown in green/white, and data from the same population one year after a severe drought (1978) are shown in purple. Note that the population size (N) was significantly reduced following the drought and that the average beak size increased in the survivors. (Redrawn with permission from Boag, P. T., and P. R. Grant. 1984. The classical case of character release: Darwin's finches *(Geospiza)* on Isla Daphne Major, Galapagos. Biological Journal of the Linnean Society 22(3):243–287.)

birds down to below 100 individuals. One of the results of the drought was that many plants on the islands were unable to flower, reproduce, and set seed. The plants that were able to reproduce were plants that made large seeds. How did this affect the ground finches? Bird size and corresponding beak size changed (i.e., through the process of natural selection) as a result of the environmental change (Figure 2.6). After the drought, the surviving birds had bigger beaks that were able to crack open the large seeds, which were the only food available. The smaller birds died because they were not able to eat the food that was present. This is a wonderful example of the process of evolution through natural selection.

The Grants continued their work with the finches for many years and were actually able to document evolutionary changes in beak sizes repeatedly over the years. During very dry years, beak size increased in response to the environment, whereas in wet years, the opposite happened, and small beaks were favored, with those birds leaving more offspring. Evolution is not always a one-directional process; the changes in the environment determine which individuals are the most fit and will leave the most offspring. Some years it is one way, other years it is another, and sometimes the environment can select for the same variations over very long periods (hundreds of thousands of years) before a volcano or an asteroid or a continental shift creates new selection criteria and the life forms affected evolve in new and interesting ways.

MACROEVOLUTION

So far we have focused on microevolution, evolution of simple characteristics and the resulting genetic response to the environment (beak size, color morph, resistance, etc.). But what about the really big *macroevolutionary* changes: how birds evolved from reptiles or how humans evolved from great apes, or how land animals evolved from fishes? How do we study these and what sorts of information can we look for? We have defined evolution as a genetic-based change in a species over time, and we have looked at single-locus genetic systems that change rapidly (on the order of one to 100 years). It may be obvious that large evolutionary changes (e.g., amphibians from fish) need millions of small microevolutionary changes and will take hundreds of thousands of years. It is not easy to change the architecture of a fish into a land-dwelling amphibian because it involves reorganization of entire organ systems and biochemical pathways. This is not a process that happens overnight. Clearly, these big, slow evolutionary events require us to employ fundamentally different approaches to study them than those just discussed.

Given the different nature of large evolutionary events, what type of evidence should we look for in order to study macroevolution? Four major lines of evidence can help us with this task: (1) the fossil record, (2) biochemical/molecular evidence, (3) structural evidence, and (4) developmental evidence. We will briefly explore each of these in turn and show how

evolutionary biologists use them to explore the world of macroevolution.

FOSSILS

Fossils and the fossil record seem to crop up a lot whenever people talk about evolution. But what exactly is a fossil? It seems that we should know something about fossils and how they are formed before we can look at fossil evidence showing macroevolutionary patterns. Fossils are preserved remains, tracks, or traces of once-living organisms, usually embedded in some type of rock. According to this definition, fossils are essentially dead things that have been preserved in rock. There have been lots and lots of dead organisms on the planet over time, so why aren't we drowning in fossils? Why doesn't everything that dies become a fossil? In other words, what has to happen in order for a fossil to form? It turns out that certain conditions have to be met for a dead organism to leave any kind of a fossil trace.

First off, a fossil is much more likely to form if the organism has some kind of hard part. These are the obvious things such as bone, teeth, claws, shells, exoskeletons, hair, feathers, scales, and so forth. If you think about it, this requirement suggests that a huge number of organisms are unlikely to ever leave a fossil trace at all. The list would include all the soft-bodied organisms (worms, sea anemones, slugs, sponges, jellyfish, fungi, many plants) and most of the small and microscopic critters (algae, protists, bacteria, viruses, etc.). In addition, this means that the soft, squishy parts of larger organisms such as brains, organs, skin, and muscle also do not leave much of a fossil record. Because we know that life evolved from simple, single-celled organisms, we might conclude that the fossil record should be fairly scarce until a time in the Earth's history when organisms got big enough and developed some hard structures that could leave a fossil record. Not too surprisingly, this is exactly what we find when we look at the fossil record; millions of years in the early history of Earth are without much fossil record, and then, in a short geological span, we find a whole host of fossils of small organisms with hard parts. In addition, this hard part requirement also means that some groups of organisms are going to leave a much better fossil record than others. For example, mollusks with their hard shells (snails, clams, mussels, chitons, etc.) have left us an enormously abundant fossil record, showing all kinds of fascinating macroevolutionary patterns.

Secondly, in order to form a fossil, a recently dead organism has to not rot. The faster something decomposes the less likely it is that a fossil is ever going to form. Decomposition begins very rapidly after death, as bacteria begin breaking down the tissue. Good places to prevent decomposition tend to be areas with little or no oxygen (most bacteria need oxygen to fuel the decomposition). Tar pits, swamps, bogs, volcanic ash flows, and the bottom of some lakes all fit the low-oxygen bill, but in general, these areas are not very common. In addition, to prevent rotting, a potential fossil needs to be covered in some way and not just left lying around. This requirement also places some big limitations on the kind of organisms that leave a fossil record. Only organisms found in certain kinds of habitats are likely to be trapped or covered and therefore to leave a fossil. Thus, a snail living in a tropical swamp is more likely to leave a fossil than is one living in the Arctic Ocean.

Finally, most things that die get eaten by something else. Even rotting is essentially the process of being eaten by bacteria and fungi, although we are discussing other processes here. How often in your hiking around in the wild do you ever find the carcasses of dead things? Probably not very often. Why? Because most things that die quickly become food for some other creature. Nature likes to recycle things like nutrients and energy, and dead tissue (be it animal or vegetable) is a good source of both of these. Most dead stuff is quickly eaten, long before fossilization can occur.

Fossilization is therefore a pretty rare event. It takes special conditions and distinctive sets of circumstances for there ever to be a fossil record of an organism. It has been estimated that only a tiny percentage of the Earth's surface at any given time has the right environmental conditions to lead to fossilization. As a result, paleontologists have concluded that it's likely that only 1 percent of all species ever found on the planet have left a fossil record. That's 1 percent of all *species,* not individual organisms. So the question should not be why are we not drowning in fossils, but rather, why do we have any fossil record at all?

Regardless of how rare fossils are, the fossil record is an enormously fruitful arena in which to look for evidence of macroevolutionary patterns. We find that there are great fossil records of many of the large evolutionary events. The evolution of birds from reptiles is well recorded in fossils and has produced one of the most famous fossils of all time, *Archaeopteryx.* This particular fossil shows an organism with many reptilian features (vertebral tail, teeth in jaw, etc.) and some surprising avian features (feathers and wings), making it a great example of evolutionary transition from reptiles to the modern lineage of birds. Other great fossil records exist for the evolution of *Homo sapiens* from the great apes, the evolution of land animals from fishes, and many fascinating evolutionary events in marine mollusks. Evolutionary biologists often generate hypotheses about how evolution works and then test the questions using the fossil record. The study of fossils continues to be a rich area of contemporary evolutionary scholarship, and the library of fossils keeps growing as new ones are uncovered.

BIOCHEMICAL/MOLECULAR EVIDENCE

One of the most exciting and dynamic areas of evolutionary research involves the use of molecular biology, biochemistry, and modern genetic analysis. Much of this work is predicated on the fact that all life on this planet originated from a common ancestor. Therefore, we should be able to find predictable patterns of relatedness at the genetic and biochemical levels. To better understand the logic behind this kind of research, let's take an example that we can all relate to: Rank the vertebrate animals in Figure 2.7A according

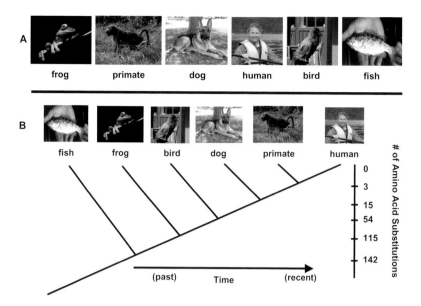

FIGURE 2.7. Six vertebrates: frog, primate (baboon), dog, human, bird, fish. Part B shows the evolutionary relationships (phylogeny) among these six animals, based on the number of subunit differences each organism's hemoglobin has when compared to the human hemoglobin molecule. Note that the evolutionary relationship (phylogeny) drawn here also includes another axis (time), which indicates the relative time in the evolutionary past when they shared a common ancestor. This is the kind of evidence that demonstrates that humans are also primates. (Modified after Goodman, M., G. W. William Moore, and G. Matsuda. 1975. Darwinian evolution in the genealogy of haemoglobin. Nature 253:603–608.)

to how similar you believe they are to human beings. List the most similar to humans first, followed by the next closest, and so forth. What does your list look like?

Here is the way most students have listed them in the past: human, monkey, dog, bird, frog, and fish. This suggests that monkeys are the most similar and fish are the most different. Now suppose you wanted to test whether there was an underlying evolutionary relationship among these organisms. What kind of pattern would you expect from their genes? Is it reasonable to expect that the two organisms that shared the most recent common ancestor would have the most similar genetic makeup and that more distantly related animals would be progressively less similar?

We can test this hypothesis by looking for patterns in the biochemistry of these organisms. Figure 2.7B shows the number of amino acid differences (amino acids are the building blocks of proteins) between human hemoglobin (the blood protein in all vertebrates that carries oxygen around in our bodies) and that of each of the other animals. Notice that monkeys and humans have the most similarity, with only three small differences between their hemoglobin molecules. Each of the other animals is progressively less similar to humans, ending with fish. Fish and humans have a whopping 142 amino acid differences between their hemoglobins. This type of graph is often called a *phylogeny* and shows the evolutionary relationships (or amount of relatedness) among organisms.

The biochemical similarity among hemoglobins strongly suggests evolutionary decent from a common ancestor. If it were not so, we would expect almost any pattern other than the one we see. Yet we find this same pattern of relatedness over and over, everywhere we look in nature, demonstrating that all life on this planet is linked and shares a common ancestor. The type

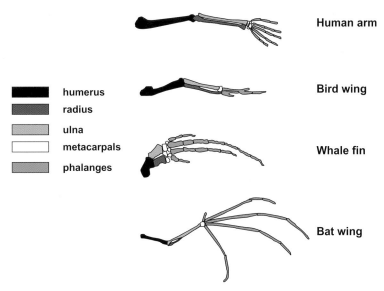

Human arm

humerus
radius
ulna
metacarpals
phalanges

Bird wing

Whale fin

Bat wing

FIGURE 2.8. The homologous bone structure of four different vertebrate limbs. Note the relative position and size differences of each bone or bone type among the four limbs, but also notice that all four share a similar basic pattern: one bone, two bones, many bones. (Figure by M. P. Marchetti.)

of logic used with this hemoglobin example is the basis for most modern molecular evolutionary research. From tracking the history of the AIDS virus to looking for patterns among endangered sea turtles to exploring the origin of the human species, molecular biology continuously elucidates macroevolutionary patterns.

STRUCTURAL EVIDENCE

A third place to look for macroevolutionary patterns, which was first used by the early biologists, involves studying similarities in physical structures among organisms. We just showed the logic used to deduce patterns from biochemical evidence. Essentially, the same logic applies here. Similarity in physical structure implies similarity in ancestry. Before we dive into an example, it is useful to define some important terms. *Homologous structures* are defined as structures that are derived from a common ancestral structure but may have different appearances and functions. *Analogous structures* are structures that resemble each other without sharing a common ancestor. One clear example of this comes from the underlying bone structure

of all vertebrate limbs. Your upper arm bone (humerus) descends from the shoulder, which articulates with a pair of forearm bones (radius and ulna), which then articulate with a whole series of small bones that make up the wrist and fingers (carpals, metacarpals, and phalanges). This basic structure (one bone connected to two bones, connected to many bones) is the same in the limbs of every single vertebrate (Figure 2.8). The wing of a bat, the flipper of a dolphin, the leg of a sheep, and the arm of a human are all considered homologous structures. Structural homology strongly implies evolutionary decent from a common ancestor. Bones in a limb do not have to be constructed this way. It would be perfectly acceptable biomechanically to have a single bone attached to a single bone attached to many bones, but that is not the way vertebrate limbs evolved. Early ancestral vertebrates that had limb bones in the above pattern left more offspring than other organisms, and all the rest of the vertebrates descended from these ancestors inherited this underlying structural organization. If vertebrates were not all descended from a common relative, then it would be very

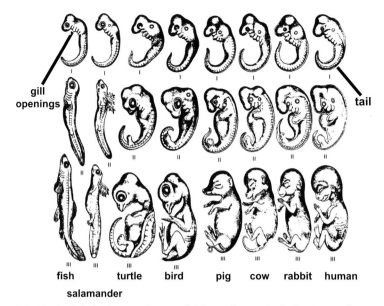

FIGURE 2.9. Early developmental stages of eight vertebrate animals showing a striking amount of similarity in general features among them, including the presence of gill slits and a post-anal tail. Similarities of physical characteristics such as these and many others are used to infer evolutionary relationships among different vertebrate groups. In the current example, all vertebrate species are believed to have originated from a common vertebrate ancestor based on these types of shared developmental traits. (Reprinted from Romanes, G. J. 1896. Darwin and after Darwin: an exposition of the Darwinian theory and a discussion of post Darwinian questions. The Open Court Publishing Company, Chicago.)

likely that some vertebrate limbs would be constructed differently. But they are not.

At this point, it is important to understand that many similar structures are not homologous, but analogous. For example the wing of a bat and the wing of a dragonfly superficially look similar (with a membrane stretched between rigid supports) and perform similar functions (flying), yet they are not the result of a common ancestry. Bats and insects are very distantly related, and their general wing structures are not descended from one another. An understanding of these types of physical patterning can be a powerful tool from which to interpret macroevolutionary patterns.

DEVELOPMENTAL EVIDENCE

Another form of physical evidence used to demonstrate macroevolution is quite similar to both biochemical and physical structural patterns but involves the patterns in growth and development

of organisms. Developmental biologists study the processes involved in organismal development, from fertilized egg to mature adult. You can actually watch a crude approximation of evolutionary history unfold in the early developmental stages of life. Again, this is due to the fact that all life is related through common ancestry. A great example can be seen in the early developmental stages of all vertebrates. Because all vertebrates are descended from a common ancestor, all vertebrates actually pass through stages in development that reveal our early ancestry (Figure 2.9). All vertebrate embryos at some point have gill slits (like fish), have post-anal tails (like many mammals), and look very similar to one another. Yet this is clearly not the only developmental pathway that can result in higher organisms. Take the example of the octopus, an extremely intelligent creature with a very high-functioning brain. Octopi can learn tasks by observation; for example, one can open a sealed jar with food in it by watching another octopus perform the operation. Yet when we look at the development of the octopus, we find that there is absolutely no similarity between it and the development of large-brained vertebrates because they do not share a recent common ancestor. Again, this type of developmental patterning is observed all over in nature and gives us strong evidence for macroevolution.

NATURAL SELECTION, EVOLUTION, AND CONSERVATION

It is important to note here that, among scientists, there is very little controversy regarding the ideas of evolution and natural selection. The theory of evolution is supported throughout the biological sciences by tens of thousands of examples and studies, with no convincing evidence to refute it. There are some non-scientists who seek to discount the certainty that scientists have in evolution by saying that it is "just a theory." This, in part, stems from the fact that, to scientists, the term theory means something quite different than it does in the culture at large. To scientists, a theory is a clearly defined set of general principles that have been mathematically described and repeatedly validated with experiments and field data. For example, physicists describe gravity and electricity with "the theory of gravity" and "the theory of electromagnetism," not because they are uncertain about the existence of gravity and electricity, but precisely because they are highly certain about the nature of these phenomena. By the same token, biologists call evolution a theory because it is a clear, powerful idea that is well supported by evidence. Few ideas in science ever have the importance, clarity, and validation to be called theories. Evolution happens continuously everywhere there is life. To deny it exists is to deny our ability to learn how the world works through observation.

So, why have we included a chapter on natural selection and evolution in an introductory text on conservation? Most importantly, it is to help you realize that every life form is constantly changing as its environment changes. Any conservation strategy has to work with the idea that we can't "freeze" species or natural areas in time in order to protect them. People interested in conservation increasingly recognize that evolutionary change is a natural process that requires tending, or else extinction results. Here are some examples:

FLORIDA PANTHER   The population of this variety of puma, confined to the lower panhandle of Florida, became so small that it lacked the genetic capacity to persist on its own, as odd mutations, such as kinked tails, became expressed. The solution to the problem was to introduce panthers from outside Florida to mate with Florida panthers, and therefore increase their genetic capacity to adapt. This is still an experiment in progress, but it seems to be working.

FISHERIES   As indicated when we talked about cod, a phenomenon observed in many fisheries is that, after humans begin to harvest a population, the size of the average fish declines. Harvesting by humans reduces the lifespan of the average fish in a population. This means that a fish is better off starting to reproduce when it is younger and smaller, because if it waits until it is older and larger, it may get harvested and not reproduce at all. A tradeoff generally exists for organisms between putting energy into their own growth and into reproduction, as indicated in the tule perch example. Heavy harvesting by humans selects for those individual fish that reproduce younger and put more energy into reproduction rather than growth, and hence results in the average fish becoming smaller in the harvested population. When the fishery stops, the size stays small, demonstrating its genetic basis.

MALARIA   Malaria was nearly eradicated in the mid-twentieth century because the mosquito species that carry the *Plasmodium* parasites were highly susceptible to the pesticide DDT, and drugs were discovered that attacked the parasites in the human bloodstream. However, natural selection favored those few individual mosquitoes that happened to be resistant to DDT and other pesticides, so that now many mosquitoes are resistant to our pesticides. Consequently, malaria is increasingly difficult to control. The evolved resistance of mosquitoes to pesticides has combined with the evolved resistance of the parasite itself to antibiotic drugs, helping to make malaria a widespread disease again; it is currently a major cause of death and illness in many tropical countries. There is some speculation that malaria may become more widespread in temperate regions such as North America with climate change, because the warm conditions necessary for the mosquitoes that carry malaria will occur at higher latitudes. Thus, the consequences of natural selection have very real implications for you, your family, and your lifestyle. There are also implications for wildlife because an outbreak of malaria is likely to result in extreme measures of mosquito control, such as the draining of wetlands or the widespread application of pesticides.

These examples demonstrate that evolutionary change can happen surprisingly fast. Unfortunately, for most organisms, we humans are changing the environment faster than they can adapt to it. This suggests that we need (1) to develop new and improved strategies for conservation, (2) to reduce the human impact on the world (for our own sake if nothing else), and (3) to create "natural" areas that are large enough in size so that natural selection has a chance to operate in ways that maintain biodiversity. At least one country, Ecuador, has officially recognized the link between conservation and evolution. The new Ecuadorian constitution (2008) states that

"Nature . . . has the right to persist, maintain, and regenerate its vital cycles, structure, and functions and its processes *in evolution*" (italics added).

CONCLUSION

We have now seen the critical role variation plays in nature and learned how differences among individuals can lead to the elegantly simple process of natural selection. We have looked at a number of examples of natural selection at work in the world and used these to fully develop a working definition of evolution. We have also compared the small-scale process of microevolution with the large scale process of macroevolution and looked at how we go about studying each of these. The information in this chapter will provide a foundation for much of the material to come, and we shall see how vitally important the ideas of variation and natural selection are for the conservation and protection of biodiversity.

FURTHER READING

Gould, S. J. 1994. Hen's teeth and horse's toes: further reflections in natural history. W.W. Norton and Co. New York. *The "popular" writings of the late Stephen J Gould, a famous evolutionary biologist at Harvard, are always both informative and entertaining.*

Weiner, J. 1994. The beak of the finch: a story of evolution in our time. Knopf. New York. *This is a nicely written account of how Peter and Rosemary Grant documented evolution in the finches of the Galapagos Islands; it is a combination detective and adventure story.*

3

Species

THE BASIC UNIT OF CONSERVATION

In order to link many of the general ideas and concepts in conservation science, you need to better understand a deceptively simple, if familiar, concept: the idea of a *species*. You have no doubt used the word (as in endangered species, extinct species, newly discovered species), but do you really have a good idea of what it means? How is a species defined? How does a species form? How do our ideas about evolution and natural selection fit with our understanding of species and conservation? We will try to answer these questions and others in this chapter.

WHAT IS A SPECIES?

Species is a term derived from Latin *(specere)* that means "a kind" or "a type." Carolus Linnaeus, in the eighteenth century, formalized the use of the term species to refer to a taxonomic unit or rank. Biologists have essentially kept this general understanding, and the classic definition of a species is the following: *a group of interbreeding individuals that is reproductively isolated from other groups.* This may sound rather obtuse, but basically it means that members of a species

breed with each other but not with members of other species.

It is important to note here that scientific species names have a particular form. All species that scientists have described are given a unique two-word Latin name. The first word is the genus and is always capitalized; the second is the species name and is never capitalized (e.g., *Homo sapiens*). A third word may be used to designate the subspecies name, and it is also never capitalized (e.g., *Canis lupis familiaris*). When writing these names, they are always either italicized (in text) or underlined (if hand written). This system was originally developed by Linnaeus in the seventeenth century, when it was thought by western Europeans that all species were the result of God's creation and were therefore unchangeable. Most of us know now that, thanks to evolution, species are constantly changing in small ways.

To explore what it means to be a species a little further, take a look at four pairs of organisms (Figure 3.1A–D). For each pair, decide if they are the same or different species, before you read further. This is a pretty easy task (regardless of

FIGURE 3.1. Four pairs of organisms for you to decide if they are the same or different species. (A) German shepherd and grey wolf, (B) Pomeranian and Great Dane, (C) broccoli and cabbage, (D) horse and donkey. (Photos by M. P. Marchetti and morguefile.com.)

the results) because we humans are remarkably good at distinguishing differences among groups of organisms. Yet among our ancestors, the basic classification system was probably something a little less complicated, such as useful organism, harmful organism, or not relevant.

PAIR #1 Are the first pair—a German Shepherd dog and a grey wolf—the same or different species? The answer is that they are the same species, but the dog is classified as a subspecies of the wolf. The dog is *Canis lupis familiaris,* and the grey wolf is actually *Canis lupis lupis.* Dogs and wolves can breed together, and you can buy a dog-wolf mix. Interestingly, these two animals were actually classified as different species until 1993, when the genetic evidence became overwhelming that they were not distinct! So if you had answered this question before the 1990s, the answer would have been that they were different species. The evidence now indicates that dogs evolved from wolves through the process of learning to live in close association with people, over the course of about 5,000 years. You may take issue with this by noting that our definition does not say anything about changing the species status of a critter or about subspecies, but rest assured that we will deal with these issues in a moment.

PAIR #2 The second pair of organisms—a Pomeranian and a Great Dane—are obviously both dogs and share the species name of *Canis lupis familiaris* with the German Shepherd. All domestic dogs fit our definition of species, in that they can breed together, although in practice this might be physically difficult for a Pomeranian and a Great Dane without some veterinary assistance. Regardless, sperm from one dog can fertilize eggs from the other and produce a puppy. New and different dog breeds such as the Labradoodle (cross between a Labrador Retriever and a Poodle) and Schnoodle (cross between a Schnauzer and a Poodle) are currently all the rage among some dog fanciers.

PAIR #3 You might think the third pair—broccoli and cabbage—are different species, especially if you like to eat one and not the other. Yet, regardless of your taste preferences, they are the same species *(Brassica oleracea),* despite the fact that they look different, taste different, and have different common names. In fact, they are also the same species as cauliflower, Brussels sprouts, kale, kohlrabi, and collard greens. Sperm from broccoli (i.e., pollen) can fertilize the flowers (i.e., eggs) from any of the other plants and vice versa, making them the same species, although the result may not be as edible.

PAIR #4 Are a horse and a donkey the same or different species? The answer is that they are different species—*Equus caballus* and *Equus asinus,* respectively. Some farm-savvy folks may note that indeed horses and donkeys can breed with one another, but the offspring (mules) are sterile, meaning that they themselves can't produce offspring. In fact, the only way to get a mule is to breed a horse and a donkey together. Mules can not make other mules; they are a reproductive dead end.

So what is going on here? We have some examples above that make sense with our definition and others that do not. How can we reconcile these contradictions? One avenue toward this end is to slightly append our definition of a species to include additional information that may help. We will now define a species as *a group of interbreeding individuals that is reproductively isolated from other groups and that can freely mate and produce fertile offspring.* By adding these two bits to our definition, we can help resolve some of the issues above. The addition of freely mating in nature allows us to include dogs and wolves as the same species because they easily interbreed when they are together. The issue with the horse and donkey is solved by the addition of the part requiring fertile offspring. The mule is not fertile, and therefore the mating of horse and donkey is not a "good mating," and they are classified as different species.

You may be feeling a bit uneasy about all the uncertainty and fuzziness around the definition of species, and that is okay. Many biologists themselves take issue with this particular definition of species (called the biological species concept, or BSC). We will explore some of the difficulties with the BSC later in this chapter, but suffice it to say that despite the quirks and ambiguities in the definition, species are a basic and fundamental unit in biology. They can be studied and quantified, and in general the concept is extremely useful. It is worth remembering, however, that the idea of a "species" is somewhat abstract and is also a human construct. We (humans) are applying a definition to the natural world that is chock full of natural variability, and so our definition does not fit everything in nature. The BSC definition applies most easily to organisms that resemble us most closely, vertebrates, and tends to become less easily applied as we try to define species of invertebrates, plants, or microorganisms.

VARIATION AND SPECIES

In the chapter on natural selection, we looked at the fossil record and saw large evolutionary changes through time. We assume that these fossils record the evolution and extinction of species. But how do new species arise? The answer lies in a topic we have already touched, namely variation. Variation is the key driver for the evolution of new species and appears most obviously in distinct groups within species. We have already encountered some of this variation in our discussion above, under the term subspecies. The domestic dog as a subspecies of wolf is a prime example. Other examples include the endangered Florida panther as a subspecies of mountain lion *(Felis concolor)*, multiple subspecies of poison arrow frogs in Central America, and the four runs of California Chinook salmon *(Oncorhynchus tshawytscha)* that we will examine later in this chapter.

Some natural variation is familiar to us under the term race, although as a term it is poorly defined and is typically used less frequently now than it was in the past. Race can refer to variation among dogs, fish, humans, and many other organisms. Some people want to lump variation in skin color and simple external features among humans under the term race. It turns out that skin color is not a particularly useful way to separate human beings, and in fact there is generally more genetic variation within so-called racial groups than is found between racial groups.

Other types of variation are given different names, depending on the type of organisms involved. For example, when domestic plants of the same species have widely differing traits, the variants are sometimes called cultivars. The broccoli and cabbage mentioned above are each different cultivars of the species *Brassica oleracea*.

Another term for plant (and sometimes animal) variation is the word *ecotype*. An ecotype refers to variation that has an environmental correlation. Let's look at an example of an ecotype using a common northern hemisphere wildflower called yarrow *(Achillea millefolium)*, which grows at many elevations in western North America (Figure 3.2). At low elevations, the plant is fairly large, with lots of leaves and an abundance of little flowers, whereas at high elevations, members of the same species are short and often have only a few leaves and tiny flowers. The interesting thing here is that many people would have a hard time identifying plants from these two elevations as the same species. They look very different, but pollen from one can be used to pollinate the other, and the resulting offspring are all fertile, so we know they are the same species. The presence of elevational ecotypes and the concept of plants being locally adapted (which means there is a genetic basis for the ecotypes) to a particular region becomes important when thinking about restoration ecology. If one wanted to restore a meadow at 9,000 feet, it would not be the best choice to get yarrow seed from plants grown at sea level. The plants from sea level are adapted to the growing conditions there and would not do well in the harsh high-altitude (alpine) environment. Restoration ecologists try instead to get local ecotypes for all their restoration work when it is possible.

All of these types of variation (subspecies, races, cultivars, ecotypes, and others) exist

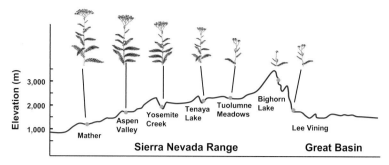

FIGURE 3.2. Distribution, elevation, and relative size of yarrow *(Achillea millefolium)* populations across a transect through the Sierra Nevada Mountains, California. Note that the relative height and flower size decrease with increasing elevation. (Redrawn with permission from Clausen, J., D. D. Keck, and W. M. Hiesey. 1948. Experimental studies on the nature of species volume III: environmental responses of climatic races of *Achillea*. Carnegie Institution of Washington Publication 581. Washington, DC.)

below the level of the species. What this means is that all the varieties can breed together and produce fertile offspring. In general, however, they don't do this. Something in nature keeps the varieties from breeding together, thereby preserving the variation we observe. But what are the forces that prevent interbreeding, and how do these same forces contribute to the formation of entirely new species? We will begin to examine these questions next.

NEW SPECIES

In order to form a new species, a group of individuals has to diverge in some heritable characteristics. This essentially means that they develop a suite of traits that make them different from other individuals of the species. The process of natural selection aids in this divergence, because the environment (predators, lack of food, environmental stress, etc.) allows a subset of individuals with favorable traits to leave more offspring until eventually they can no longer breed with the original group. How can this happen? What forces can cause populations to become less similar through time and eventually lead to the formation of a new species? The answers lie in two general categories: *pre-zygotic isolating mechanisms* and *post-zygotic isolating mechanisms*. The names sound daunting, but they are relatively easy to understand if we remember the definition of zygote from

high school biology. When an egg and sperm (also called gametes) join during reproduction, the resulting diploid cell is called a zygote. So these two types of forces are things that isolate a group of individuals before a zygote is formed (pre-zygotic) and after a zygote has formed (post-zygotic). Let's look at each of these in turn.

Pre-zygotic isolating mechanisms are factors that come into play before fertilization (mating) that allow some individuals of a species to become divergent or different through time. We are going to examine six of them.

The first factor is probably the easiest to understand and is *geographic separation*. The idea is that some geographic barrier (river, mountain range, ocean, etc.) isolates two populations of a species. Over time, the two groups diverge in their characteristics until they can no longer interbreed. A good example of this occurs with oak trees in the genus *Quercus*. If we examined three of the many species of oak trees (*Q. rober,* English oak found across Europe; *Q. lobata,* California valley oak found in the Central Valley of California; and *Q. dumosa,* coastal scrub oak found in the coastal region of California), we would see that they are all very similar in their general characteristics (i.e., leaf shape and size). In fact, they could cross-pollinate each other except for the geographic barriers that prevent their reproduction. An entire ocean and continent separates the English oak from the other two, effectively cutting off any chance of pollination. The two California species are also separated, this time by the California Coast Range, mountains that have peaks over 5,000 feet high. Oak trees have a hard time crossing mountaintops, and as a result, the two species are isolated from one another. Distance and physical barriers are effective means for preventing reproduction and allowing speciation to occur.

The second factor is called *ecological or habitat separation*. Here, the idea is that two populations can occur in the same geographic region without any physical barriers separating them, if they use different habitats. An interesting example of this comes from the Indian subcontinent, where we can find two large carnivorous cats, the Asiatic lion *(Panthera leo)* and the Bengal tiger *(Panthera tigris)*. Lions live in the open savannahs and use stealth and speed to hunt their prey (note their coloration), whereas tigers live in the forest and use camouflage and surprise to hunt their prey (note the stripes). In nature, you would not find lions in the jungle, just as you would not find tigers out in the open, yet these two animals can breed if they are forced together in a zoo. The offspring of this mating is variously called a tigon or a liger, but it is generally sterile. The differing habitats, ecology, and natural behavior patterns prevent the mating between these two animals.

The third factor is called *behavioral separation* and results from differences in the ways two groups communicate with each other. Many species communicate through gestures, songs, or chemical means, all of which are aspects of their behavior. If two groups are located in the same geographic area and reside in the same habitat, differences in their behavior can prevent them from interbreeding. A good example of this comes from fruit flies on the islands of Hawaii. Fruit flies all generally eat the same things and therefore are often near or in close contact with each other, yet there are many species of fruit flies. One of the forces that keeps the species isolated is reproductive behavior. For example, two species of fruit fly, *Drosophila silvestris* and *Drosophila heteroneura,* overlap in

their distribution and eat the same food but have very different mating rituals. In *D. silvestris* the male approaches the female from the rear and then drums on her abdomen with his front legs to initiate mating. In *D. heteroneura* the male approaches the female from the front with his wings extended to initiate mating. What happens if a male *D. silvestris* approaches a female *D. heteroneura*? The female does not recognize the "language" that the male is using and ignores his advances. Much the same thing might happen if an English speaker went into a village to find a date where only the !Kung dialect was spoken (as shown in the movie *The Gods Must Be Crazy*). Not much is likely to happen. Aspects of their behavior prevent mating from ever occurring.

A fourth mechanism is called *temporal or seasonal separation* and has to do with timing. Suppose that two groups of organisms live in the same habitat and essentially speak the same behavioral language but that the timing of their reproductive efforts is off by months. It is not likely that individuals in the two groups will ever interbreed. An example of this comes from the western United States, where two species of trout, brook trout *(Salvelinus fontinalis)* and rainbow trout *(Oncorhynchus mykiss),* have been widely introduced into many streams and rivers. Brook trout are fall spawners, typically breeding from September to early January, whereas resident rainbow trout are spring spawners, typically breeding from February to June. In a hatchery, these two species can be interbred to produce a sterile hybrid, but in nature this does not happen because the reproductive timing of the two species does not match up.

The fifth mechanism is called *mechanical separation* and may make some readers chuckle a bit. The idea here is that two groups may reside in the same habitat, speak the same language, and have similar reproductive timing, but when they get down to the actual mating process, their reproductive organs are not compatible (i.e., they don't fit together). This happens often in insects, as we can see with the insect genitalia in Figure 3.3. These are scanning electron

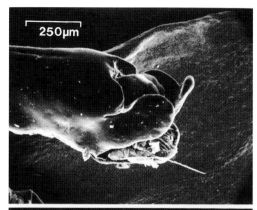

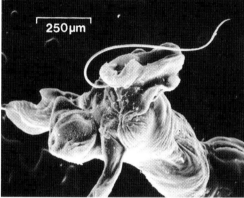

FIGURE 3.3. Scanning electron microscope pictures of male genitalia from two different species of dragonfly. Note the very different shapes, which make it impossible for mating between the two species to occur. (Reprinted with permission from Krebs, J. R., and N. B. Davies. 1987. An introduction to behavioral ecology, second edition. Blackwell Publishing. Oxford.)

happens that gametes (egg or sperm) from one group are not able to fuse with gametes from another group. One version of this happens all the time with plants. Wind-dispersed pollen (male plant gametes) blows around indiscriminately and lands on the female portions of many different species of flower. Yet it is only the correct species of pollen that can actually fertilize the flower; all other pollen is biochemically rejected and won't cross-pollinate the wrong species.

All six of these forces act before a zygote (embryo) is formed, yet there is also a suite of other forces that can act after fertilization, called *post-zygotic mechanisms*. This can happen in a number of ways. For example, when two unlike critters get together, gametes are exchanged, and fertilization occurs, but they still might not produce good offspring. The embryo of a mating between two unlike organisms can fail to develop properly and die prior to birth. Or the embryo may fully develop past birth, but the resulting offspring may be inherently weak or unfit, and thus may fail to survive in nature. Finally, a hybrid offspring may survive to adulthood, but as we have seen with the mule, it may be sterile and not able to reproduce. A classic example of this occurs among freshwater sunfishes, commonly caught by kids in the lakes and streams of North America. Males of a number of these species often mate with females of other species, but the results are invariably sterile and look like perfect intermediates between the two parental species. These three forces act after a zygote has formed and can prevent proper reproduction from occurring.

All of these mechanisms (pre- and post-zygotic) can arise in populations that are diverging and prevent individuals from one group from mating successfully with individuals from another population. Species formation is a continuous process and takes many generations. If two species are diverging (say geographically) and they somehow come back into contact, the mating that would result can make the differences between populations eventually disappear. It turns out that reproductive isolation is

micrograph pictures of the male copulatory organs of various dragonfly species, which are widely different in shape and size. The female orifice that would receive one type is not likely to fit the other. In fact, the reproductive organs of insects are so widely divergent that lepidopterists (aka butterfly-ologists) actually use them to help correctly identify butterfly species. In short, sometimes the mechanics of reproduction prevent the act of mating from occurring.

The final pre-zygotic mechanism is called *gamete isolation* and is essentially the last barrier preventing some taxa from reproduction. All other forces being equal (i.e., same habitat, anguage, timing, mechanics, etc.), it sometimes the key to forming a new species. But how long does species formation take?

RATES OF SPECIATION

There is no single answer to the question of how long it takes for a new species to evolve, because different organisms evolve at different rates. Simple organisms such as bacteria or viruses evolve rather quickly, whereas large, very complex organisms such as great whales and redwood trees evolve rather slowly. There is a dichotomy of thought among biologists as to the rate or tempo at which organisms evolve, with essentially two sides to the issue. One camp suggests that speciation is a slow, even-paced process and is given the name of gradualism. This is essentially the way Darwin thought evolution proceeds. The top portion of Figure 3.4 graphically depicts how this might happen. Two populations of a species slowly accumulate small genetic differences over a long time period. This is often believed to occur with one or more of the pre-zygotic isolating mechanisms at work keeping the two populations apart. Eventually, the divergence is great enough between the two populations and they are no longer able to mate. We can then declare them to be two different species.

The second camp is a departure from gradualism and was championed by Niles Eldredge and the well-known Harvard professor Stephen J. Gould. Eldredge and Gould both studied fossils and were convinced that the fossil record revealed a different evolutionary story than was told by gradualism. According to the fossil record, most organisms remain static and unchanging for long periods of time, followed by the very abrupt and rapid evolution of a host of new species over very short time periods. The geologic record shows that these major spurts of speciation are often strongly correlated with major environmental changes such as asteroid impacts, volcanic eruptions, and ice ages. This different idea for the pace of speciation was given the name *punctuated equilibrium* to emphasize both the rapid nature of speciation

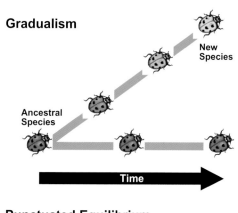

FIGURE 3.4. Cartoon representations of two different views on the pace, or speed, of speciation and evolutionary change. One view is that evolution works slowly, with a gradual accumulation of small changes—hence the term gradualism. Another is that the fossil record shows long periods of stasis or no change followed by rapid speciation events, a process called punctuated equilibrium. (Figure by M. P. Marchetti.)

and the long-term periods of stasis and equilibrium between the bursts of evolutionary diversification. The bottom portion of Figure 3.4 depicts how this might look.

The dichotomy of thought over this issue has not been resolved, but this may be due to the fact that both ideas may actually be correct. It is possible that, for some organisms, evolution may proceed slowly and follow the path of gradualism, whereas for others, speciation may be fueled by environmental catastrophe. It is also important to keep in mind that what appears to be a short time in the rocks and fossil record may, in fact, be what we might consider a relatively long time (1,000 or 10,000 years). So it is possible that punctuated events do occur, but that the "rapid" evolution following them takes a good amount of time to occur. Either way, this is a fascinating area of evolutionary biology, and there are many good books out there written on this topic for non-biologists.

PROBLEMS WITH THE BIOLOGICAL SPECIES CONCEPT

At this point, we have spent a number of pages describing in detail the biological species concept and the evolutionary processes leading to speciation. This is all well and good, but unfortunately we have created a bit of a convenient fiction with the biological species concept (BSC). Yes, in practice, the BSC provides an extremely useful working definition and follows a fairly intuitive set of logical ideas, but there are some very practical problems that crop up when applying the BSC broadly across the natural world. Perhaps you have already thought of some issues demonstrating that the BSC, as described above, does not work very well. Let's tackle three of the larger ones.

We can approach one place the BSC breaks down by examining an assumption hidden in the wording of the definition. We said that a species is a group of interbreeding individuals that is reproductively isolated from other groups and that can freely mate and produce fertile offspring. This framework assumes that we have two organisms that mate by exchanging gametes, which is fine for all sexually reproducing organisms on the planet. But this wording actually leaves out a gigantic number of species, which, by their very nature, do not ever sexually reproduce. Many organisms instead reproduce by asexual reproduction. This involves only one individual splitting or budding to create a daughter cell or exact replicate. There is no exchange of gametes, and therefore there is no chance of reproductive isolation. The BSC makes it extremely challenging for us to define any of the asexual organisms as species because we can not tell where one species ends and another begins. So the BSC does not apply very well to asexual species.

A second difficulty with the BSC has to do with fossil species. The phenomenal history of biodiversity on this planet is recorded largely in the fossil record. Yet when we try to apply the BSC to fossil species, we run into a bit of a problem: our definition says we can call two groups by different species names if they no longer can interbreed. Yet how do we decide if two ancient fossil "species" were able to successfully exchange gametes? We basically have no idea and no possibility of ever finding out the truth of the situation. Fossil "species," therefore, will always have to remain presumptive species because of this inherent difficulty.

Our final difficulty with the biological species concept deals with the ever-present issue of variation and a very practical conservation application. We saw very clearly in the beginning of this chapter that natural variation exists below the level of species and is given various names (varieties, races, etc.). In a lot of species, we have sufficient variation to describe unique subspecies. Many of these subspecies do not overlap in either time or space (and therefore do not breed), and as a result, the variation at the subspecies level is preserved. If they were placed in the same area or allowed to breed at the same time, these groups could exchange gametes successfully. Yet they do not. This becomes a conservation problem when we decide to protect or preserve a species. Does this mean that we need to protect all the different subspecies and other varieties as well?

To make this more concrete, let's look at the example of Chinook salmon (*Oncorhynchus tshawytscha*) in central California. In California's Central Valley, there are four distinct runs of Chinook salmon, each of which is genetically different from the others. All four runs have a different life history pattern and follow a different calendar for the major events (i.e., migration, spawning, etc.) during their lifetime (Figure 3.5), yet each of these runs is still capable of interbreeding with all the others. However, they generally do not, because of the variation in their life history. The problem comes when we find out that two

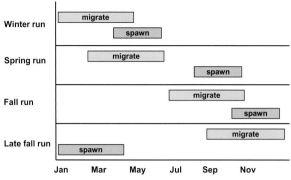

FIGURE 3.5. Four distinct populations, or runs, of Chinook salmon (*Oncorhynchus tshawytscha*) are found in California streams and rivers, each with its own unique life history pattern. For example, adults of each run vary seasonally with regard to the time of year when they enter fresh water, migrate upstream, and spawn, yet all four runs are still the same species and can potentially interbreed. Unfortunately, two of the runs are critically endangered (winter and spring runs) and are listed as endangered under the Endangered Species Act (ESA), whereas the other populations are in decline. (Drawn from data in Moyle, P. B. 2002. Inland fishes of California. University of California Press, Berkeley.)

of these runs (spring run and winter run) are in extreme danger of going extinct. If we want to protect Chinook salmon, do we need to consider all of these runs? If we protect only one of the runs, have we succeeded in protecting the species? Or could we even let all the runs go extinct because Chinook salmon spawn in dozens of other rivers from Oregon up to Alaska. In other words, does the definition of species for conservation purposes include the natural variation that exists below the level of species? Under the BSC, all four of these runs are lumped together, which implies that we need only to protect one of the four runs to protect the species. The BSC as it is defined may actually hinder conservation efforts aimed at the natural variation below the species level, which we have seen is extremely important as it is the grist for the mill of evolution. Thankfully, the architects of the Endangered Species Act took this into account when they built this groundbreaking piece of legislation, counting "distinct population segments" (such as these genetically distinct salmon runs) as species, as we will see in a later chapter. The four runs of salmon can be treated as separate species

because, for the most part, they behave like separate species under the BSC. Presumably, given enough time in a fairly constant environment, each run would evolve into a full, undisputed species.

CONCLUSION

The biological species concept is not perfect because there are many ways organisms can perpetuate themselves and their "kind." The issues discussed above are but three of many challenges scientists have with the BSC. Other definitions of species have been proposed, and many of them are either richly biotechnical or densely philosophical in nature, so we have chosen not to discuss them here. Regardless, it is important to remember that species are the basic units of biology and conservation because they are recognizable units of evolution. Understanding what constitutes a species and how natural variation interacts with environmental change to produce and change species is something that we will return to many times in the coming chapters. Indeed, a basic goal of conservation biology is to manage the environment in ways that allow species to continue to evolve and adapt to changing conditions, as they always have.

FURTHER READING

Darwin, C. 1859. On the origin of species: by means of natural selection, or the preservation of favored races in the struggle for life. Dover Publications. Mineola, NY. *Although rather long and detailed, this book lays out in startling detail Darwin's argument for evolution through natural selection and is worth the time to read.*

Gould, S. J., and S. Rose. 2007. The richness of life: the essential Stephen Jay Gould. W.W. Norton and Co. New York. *Great introduction to Gould's writing and ideas. He has a stellar ability to go from the particular to the general—to use nature's myriad examples to illustrate the richness of evolutionary theory.*

Stockwell, C. A., A. P. Hendry, and M. T. Kinnison. 2003. Contemporary evolution meets conservation biology. Trends in Evolution and Ecology 18:94–101. *This is an example of a conservation article in a scientific journal that publishes short review papers that are often quite accessible to someone with only moderate training in biology.*

4

Climate and Global Patterns of Distribution

The number of species residing on planet Earth is staggering. Current estimates suggest that the number is somewhere in the neighborhood of five to 30 million. And the number of distinct evolutionary units below the species level is probably an order of magnitude higher (see Chapter 3). Of course, all of these species aren't found in all parts of the globe, and most are quite restricted in their distribution. The enormous diversity of life around the globe is something we are really just beginning to appreciate, as shown so well by television's *Planet Earth* series and the program's accompanying book. The objective of this chapter is to introduce you to the natural environmental factors that influence the contemporary distribution of all these species.

BIOGEOGRAPHY

The study of the geographical distribution of life is called biogeography and is useful because certain patterns in the distribution of species follow some simple rules. For example, monkeys and their relatives are generally found in tropical areas, and kangaroos are limited to Australia and some nearby islands. Elephants occur in Africa and parts of southern Asia, whereas polar bears and walruses are found only in Arctic areas of northern North America and Asia. Based on the distribution of species and groups of species, we can categorize the world as consisting of a series of biological regions, or *biomes* (Figure 4.1). Biomes are regions largely defined in terms of general climate and vegetation patterns, so they do not have hard and fixed boundaries. But understanding the factors that produce the climates and biomes of the world can provide some insight into biodiversity patterns worldwide.

Two general classes of factors have led to the observed distribution of life: historical factors and ecological factors. Historical factors include events such as advances and retreats of continental glaciers, lifting of mountains, formation of islands, and slow shifting of the continents across the surface of the globe. These slow factors constitute a major area of scientific inquiry,

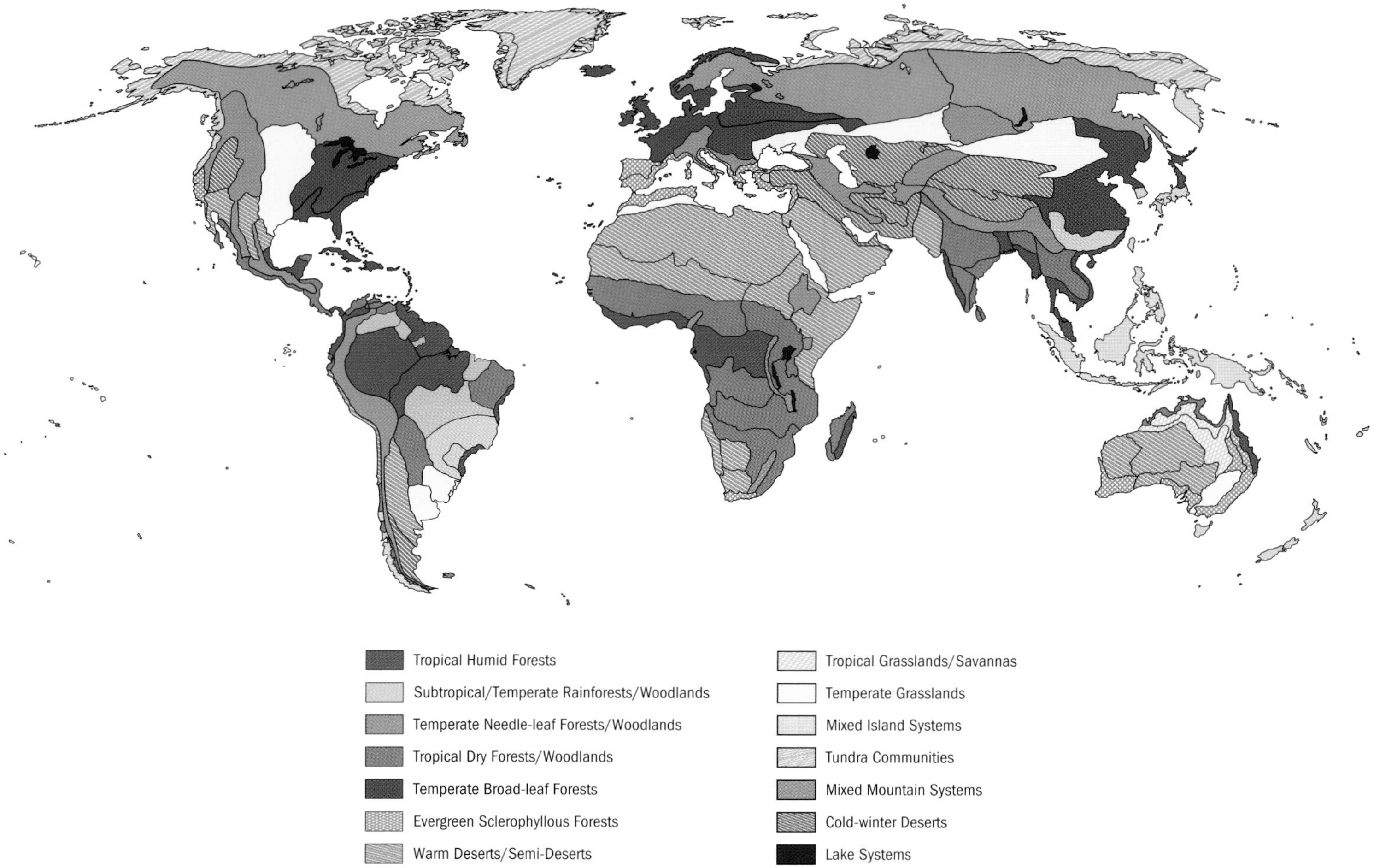
Tropical Humid Forests
Subtropical/Temperate Rainforests/Woodlands
Temperate Needle-leaf Forests/Woodlands
Tropical Dry Forests/Woodlands
Temperate Broad-leaf Forests
Evergreen Sclerophyllous Forests
Warm Deserts/Semi-Deserts
Tropical Grasslands/Savannas
Temperate Grasslands
Mixed Island Systems
Tundra Communities
Mixed Mountain Systems
Cold-winter Deserts
Lake Systems

FIGURE 4.1. Map showing the biomes of the world. (Reprinted with permission from World Conservation Monitoring Center. 1992. Global biodiversity: status of the Earth's living resources. Chapman and Hall. London.)

but in this chapter, we focus on factors that have a more immediate effect on animal and plant distribution, ecological factors. Ecological factors include such things as the timing and distribution of solar radiation and rainfall, the influence of latitude, the proximity to large water bodies, and the effect of elevation.

SOLAR RADIATION

You may remember from basic science that our planet spins like a top that is tilted slightly in relation to the sun. To be precise, the Earth is tilted by 23.5°. This simple fact actually has profound implications for much of the life on the planet. As the Earth rotates around the sun, this tilt is retained, and the sun appears to shift north and south with the changing seasons. In the summer in the northern hemisphere, the sun is located relatively far north. But in the fall, the sun gradually appears to shift further south, and in the spring, the sun appears to slowly shift northward again. This endless progression results in the seasons that characterize life in many parts of the world. Thus, when it is summer in the United States, it is winter in Australia, and vice versa.

Since ancient times, humans have marked the movements of the sun with various names and ceremonies. The summer solstice occurs on June 22 and marks the day when the sun has made its greatest progression northward. The winter solstice, on December 22, marks its southernmost progression. The halfway points are also marked by the autumnal (fall) equinox (September 22) and the vernal (spring) equinox (March 22). Historically, the year was believed to begin after the autumnal equinox, when the good times were finished and the long, cold nights of winter were approaching.

Also related to the movements of the sun are specific latitudinal markers. Of course, the equator (0° latitude) marks the Earth's midpoint. At both the vernal and autumnal equinoxes, the sun lies directly over the equator. On the summer solstice, when the sun is at its most northern extent, it lies directly over the Tropic

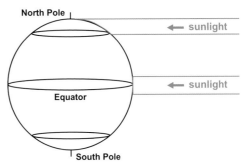

FIGURE 4.2. Because the Earth is a sphere, sunlight strikes the equator more directly (i.e., at right angles) than at the poles, and the equator therefore gets more direct sunlight and more solar energy than the poles. (Figure by M. P. Marchetti.)

of Cancer (23.5° N latitude), and on the winter solstice, it lies directly over the Tropic of Capricorn (23.5° S latitude). The tropical latitudes are often defined as the band of the Earth that lies between these two lines. When the sun moves far enough south, its light no longer shines on areas in the extreme north. The Arctic Circle is located at 67.5° N latitude and marks the latitude above which the sun never rises in the longest days of winter. Of course, in the summer, the reverse is true—the sun never sets above this latitude. And conversely, there is also the Antarctic Circle, which has the reverse seasonal characteristics. These simple observations play a large role in determining the global distribution of biomes.

Solar energy forms the basis of most of life on Earth, as it is central to photosynthesis. Photosynthesis is the biochemical process in which plants convert oxygen and carbon dioxide into sugars, which are then combined to form more complex molecules such as carbohydrates, proteins, and lipids. The Earth essentially receives the same amount of energy all the time (what astrophysicists call the solar constant). However, because the Earth is round, more solar energy hits the equator than the poles (Figure 4.2). Additionally, sunlight must pass through a greater amount of atmosphere at higher latitudes, further decreasing the amount of solar energy the poles receive (Figure 4.2). One consequence of this is that higher latitudes (near the poles) are generally cooler than lower latitudes (near the equator).

GLOBAL CLIMATE

Now we are ready to evaluate the consequences of the Earth's tilt, its round shape, and the differences in solar energy. We know that, on average, the sun spends more time directly above the equator than any other part of the Earth, and therefore that the equator receives more solar radiation than do the higher latitudes (see Figure 4.3). If we combine these facts with four relatively simple observations, we will be able to explain why it rains so much in tropical rainforests, why we have deserts around the globe, and even why France makes good wine. The four observations are ones that most of us have made on our own, and we will therefore approach them by asking four sets of simple questions.

FIRST QUESTION Which weighs more, hot air or cold air? And how do you know?

The answer is easy. Hot air weighs less, and we know this because hot air rises (think about a hot air balloon).

SECOND QUESTION Is there more air pressure at the Earth's surface or at 20,000 feet? Again, how do you know?

The answer here is also relatively easy, especially if we think about flying in an airplane. The flight attendants always warn that oxygen masks will drop if the cabin loses pressure, meaning that the inside of an airplane is pressurized and that outside at 20,000 feet elevation there is less pressure. So there is more air pressure at the Earth's surface and less as you go up in elevation.

THIRD QUESTION If you compress air, does it heat up or cool down? And how do you know?

For the answer, think about a bike or car tire for a minute. If you have ever pumped up a bike tire with a hand pump, you may recall that as you put more air in the tire (i.e., more pressure), the tire stem and the pump actually get hot (try it if you doubt it). Or, if you don't have that experience, perhaps you have let air out of a tire (i.e.,

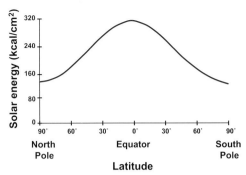

FIGURE 4.3. Graph showing the total annual amount of solar energy received at varying latitudes around the globe. Note the large peak at and near the equator. (Modified after figure from Pianka, E. R. 1988. Evolutionary ecology, fourth edition. Harper and Row Publishing. New York.)

air under less pressure, expanding) and felt the coolness as air expands. Either way, we can realize that compression (more pressure) heats air up, and expansion (less pressure) cools air down.

FINAL QUESTION Which holds more water, hot air or cold air? And how do you know?

The answer to this lies in something we all have done at some point and that is breathe onto a window or piece of glass. What happens? You get water droplets (condensation) forming on the glass. Why? Our breath (with a load of water vapor) is coming out of our bodies at about 98°F and hitting a relatively cold surface, such as a piece of glass, causing the water vapor to condense. The air (breath) in contact with the cold surface (glass) can't hold the same amount of water that the hot air coming out of our bodies can hold. So hot air holds more water than cold air.

Now, armed with these four simple observations combined with the information on solar radiation, we have enough information to tackle the basics of global climate.

TROPICAL RAINFORESTS

Let's start our global climate tour in the tropics and try to determine why we have gigantic rainforests around the equator. The answer lies in the set of facts we have just accrued. The strong sunlight at the equator heats up

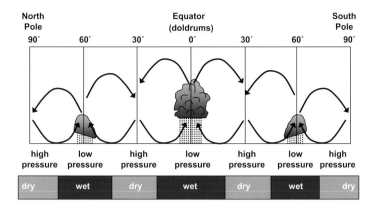

FIGURE 4.4. Global wind and rainfall patterns by latitude driven by solar energy. The large cyclical nature of these global weather cells causes patterns in both atmospheric pressure and general climate. (Figure by M. P. Marchetti.)

the air, which acts like a sponge and pulls in a ton of gaseous (evaporated) water. This hot, water-laden air then rises. As it rises, it is under less pressure (at higher altitudes), so it expands and cools (think bike tire). As it cools, the air can hold less moisture (think breath on glass), and so you get condensation, the formation of clouds, and eventually large amounts of rainfall (Figure 4.4). We already assumed that it rains a lot in rainforests, so it should come as no surprise to find out that abundant equatorial sunshine combined with a little bit of common-sense physics explains this rainy pattern. Now we know why it rains a lot in Brazil.

You have probably heard meteorologists talk about high-pressure zones and low-pressure zones. High atmospheric pressure is generally associated with good weather, whereas low pressure is associated with cloudy or stormy weather. The reason for this is also relatively straightforward. Air consists of molecules of oxygen, nitrogen, carbon dioxide, and other gases, and thus air has mass. Anything that has mass is subject to the pull of gravity, so even a mass of air puts some pressure on the surface of the Earth. However, when air rises, the pressure it exerts is reduced, and we refer to this as a region of low pressure (somewhat counterintuitive, we know). So, in areas of low pressure (such as the tropics), air is rising and generally follows the pattern we just described—air rises, air cools, water condenses to form clouds, and we often get rain.

Now we'll discuss a portion of the Earth that is characterized by just the opposite influence—high atmospheric pressure.

DESERTS

Air rising in the tropics can't rise forever. If it did, then it would escape the Earth entirely, and we would be left without an atmosphere and would have a lifeless planet. But remember that air does have mass, and thus it is subject to the pull of gravity. As a result, when air rises it expands and cools, and as it cools, it begins to become heavier again (remember hot air rises, cool air sinks). However, the hot air rising from below continues to rise, so the cool air has to go somewhere, and so at a high altitude, it diverts toward the north or south. As it moves north (or south), it continues cooling and begins falling back toward the surface of the Earth, particularly around 30° latitude (north or south) (Figure 4.4). As the air drops in altitude, it begins warming again (more pressure).

But remember that this air has already lost most of its water on its way up in the tropics (i.e., through rainfall on rainforests). So we have cool, very dry air coming down to the Earth's surface. As it gets near the surface, it is put under more pressure (think airplanes again) and warms (think bike tire), and this warm, dry air acts as a sponge, literally drawing moisture from the environment around it. This air coming down around 30° latitude sucks moisture from the landscape, making it very dry, and as

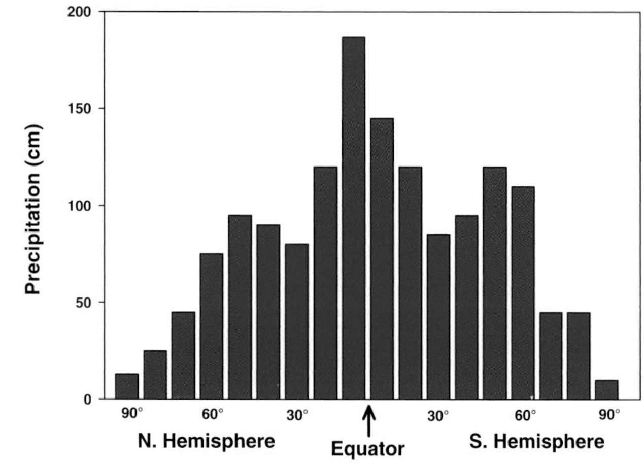

FIGURE 4.5. Average annual precipitation totals by latitude. Note the peaks around the equator and 60° latitude and the depressions around 30° and 90°. (Modified after figure from Pianka, E. R. 1988. Evolutionary ecology, fourth edition. Harper and Row Publishing. New York.)

a result, we get a desert. If you refer back to the map of global biomes (Figure 4.1), you will see that at 30° N and S latitude, we find the vast majority of the world's deserts!

To complete one cycle, the air that descends at around 30° latitude then moves north or south along the Earth's surface, making wind (Figure 4.4). Air that moves toward the equator becomes captured in the cycle of rising tropical air that we discussed above. This cycle was initially described by a meteorologist named Hadley, and the "cell" of air movement from the equator to 30° is known as the *Hadley cell*. There are two other cells (see Figure 4.4), although we won't worry about their names.

All the air coming down to the Earth's surface at 30° latitude does not get deflected toward the equator; some moves along the ground toward the poles from the 30° region, and then rises again at about 60° latitude (Figure 4.4). This air heats up, rises, cools, forms clouds, and spreads both north and south. If you have ever been on the Olympic Peninsula of Washington state, then you have experienced the mossy rainforest that lies close to 60°. In fact, we find that at around 60° latitude, we get another band of very rainy areas, mainly on the coasts, that are sometimes called the temperate rainforests.

Finally, at the poles, we have another mass of air that descends toward the Earth's surface, similar to that which causes deserts at 30° latitude (Figure 4.4). In fact, we get very little precipitation at the poles, and they are often referred

to as polar deserts. This may sound counterintuitive, but remember that it's generally pretty cold at the poles, and so whatever precipitation does fall (i.e., snow) remains there for a long time and doesn't go away. So it's cold at the poles, but they do not actually receive much in the way of precipitation, very similar to a desert (see Figure 4.5). Note on this figure that the peaks of precipitation occur at the equator and at 60°, and the lows occur around 30°. Thus, there is a generally predictable pattern of air movement over the Earth, and it is largely responsible for the distribution of rainforests and deserts.

HORSE LATITUDES AND DOLDRUMS

We have intentionally oversimplified this subject. Imagine that you are standing at the North Pole. If you think about it, you are standing on the axis of where the Earth rotates, so you would slowly turn a complete circle in one day. If you walked south a little bit and stood in one place, you would find that over 24 hours you would move a certain distance. That distance would depend upon how far from the pole you walked. The circumference of the Earth is about 4,000 kilometers at the equator, so if you stood on the equator, you would travel about 4,000 kilometers every 24 hours! This may seem a little esoteric, but imagine instead that you were in a hot air balloon, and you flew north from the equator with the northbound air in the Hadley cell. You would start your trip with a certain

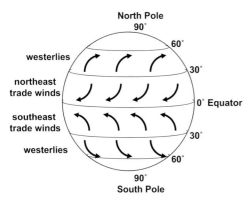

North Pole
90°

60°
westerlies
30°
northeast
trade winds
0° Equator
southeast
trade winds
30°
westerlies
60°
90°
South Pole

FIGURE 4.6. Patterns of wind shift due to the Earth's rotation (Coriolis effect). (Figure by M. P. Marchetti.)

velocity—about 4,000 kilometers per day—but as you moved north, the ground below you would be traveling east (because its spinning) at a slower and slower rate (remember, at the pole it takes 24 hours to move a much smaller distance, meaning you are moving very slowly). As this happens, another bit of physics comes into play and is referred to as the *law of conservation of momentum*. It is a mouthful, but all it means is that you don't give up the 4,000-kilometers-per-day velocity as you move north. Instead, you keep moving east at this rate. However, the Earth beneath you moves east more slowly. Thus, rather than traveling north, you will appear to begin veering toward the east, which is toward your right (Figure 4.6). The further north you travel, the more rapidly you will veer east. This same thing happens to the air in the Hadley cell; as it moves north, it shifts eastward.

The reciprocal situation would involve an air mass moving toward the equator. As the air mass approaches the equator, the ground beneath it begins moving faster, and the air mass appears to veer westward, which again is toward the right of the direction the air mass is traveling. This intriguing pattern is called the *Coriolis effect*. The important thing to understand is that, in the northern hemisphere, the Coriolis effect results in air shifting to the right when it moves to either higher or lower latitudes. The reverse is true in the southern hemisphere—air masses shift to the left.

Now, recall that air generally rises at the equator, descends at about 30°, and then moves either south toward the equator or north toward the rising air at about 60°. Air moving south from 30° will veer westward, whereas air moving north from 30° will veer eastward (in both cases, the air veers to the right of its line of travel). Many climatic features were given names during the days when sailing ships surveyed the Earth. When these ships traveled from Europe toward the New World, they would travel south to intercept westward flowing winds. As a result, these became known as the trade winds, to reflect their importance in commerce. These ships would return to Europe by a northern route, capturing eastward winds that were subsequently called "westerlies." Of course, the southern hemisphere has a similar set of winds at comparable latitudes.

In the vicinity of the equator, however, air is generally moving up, and there is not so much lateral movement. Here, winds are often poor, and many early sailing ships became stranded for weeks or more. These areas are called the *doldrums*. Similarly, at about 30°, air is descending but not providing much lateral motion. Many ships became stranded here as well and were sometimes forced to jettison or eat their cargo, including horses—thus the term *"horse latitudes."* It's interesting to note that the quickest sailing route from Europe to North America was not the most direct. Because of the Coriolis effect, ships could make this trip most rapidly by sailing south from Europe, then west with the trade winds; this brought ships to the Caribbean or southern North America, and they would then travel north along the coast of North America to reach ports such as Boston and New York.

OCEAN PATTERNS

Wind drives the surface waters of the oceans, and therefore is largely responsible for the major patterns of oceanic circulation (Figure 4.7). This results in the clockwise circulation patterns observed in the northern hemisphere and the counterclockwise patterns in the southern

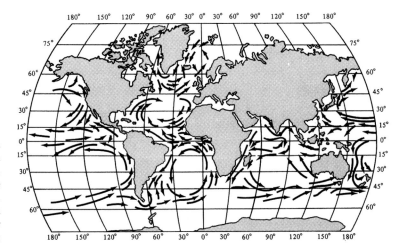

FIGURE 4.7. Ocean circulation patterns driven by global weather patterns. (Modified after figure in Pianka, E. R. 1988. Evolutionary ecology, fourth edition. Harper and Row Publishing. New York.)

hemisphere. However, there is one very important exception to this, and it results in a particularly important phenomenon called upwelling. We will use the west coasts of California and South America to exemplify this.

If we first look at currents along the coast of California, we note that they largely run parallel to the coast moving southward. As this ocean water moves south, it is subjected to the Coriolis effect, and it is shifted to the right (which is toward the west), resulting in surface waters being moved offshore (Figure 4.7). This water has to be replaced, however, and the water that replaces it wells up from deeper levels off the continental shelf and is referred to as coastal upwelling.

This deeper water is colder than the surface water and very rich in nutrients, because it brings them from the deeper ocean. This cold, nutrient-rich water is great for marine life, and as a result, upwelling areas along the California coast are very productive and full of marine life. Phytoplankton (single-celled marine plants) are fertilized by the nutrients brought up with upwelling waters and thus occur at very high concentrations here. Phytoplankton feed zooplankton (tiny invertebrates) and fish, which in turn feed larger fish, birds, and marine mammals, creating a dynamic and rich marine community. The productive salmon fisheries off the west coast of North America are also the result

of upwelling. A similar situation exists off the coast of South America. The huge anchoveta fishery of Peru is a direct result of this rich upwelling of nutrients, as we will explore in a later chapter.

LOCAL VARIABILITY IN CLIMATE

RAIN SHADOW

If you looked very carefully at Figure 4.1, you may have noticed that not all deserts occur at exactly 30° latitude. In fact, if you have ever traveled in Nevada, Utah, or eastern Oregon, you may recall that this region is very desert-like. Indeed, this region is called the Great Basin Desert, one of four major deserts in North America (the others are the Sonoran, Chihuahuan, and Mojave). But the Great Basin Desert extends from southern Nevada (about 35°) to southern Alberta (about 55° N) and is not really the product of the air movements that we examined above. Instead, the Great Basin Desert lies in a latitudinal band that has westerly winds blowing across the Pacific Ocean, picking up loads of moisture and then moving across California. Yet between California and Nevada, there is a major barrier, the 10,000–14,000-foot mountain range of the Sierra Nevada. Wind has a hard time blowing through solid rock, so instead, it is forced to blow up and over the mountains.

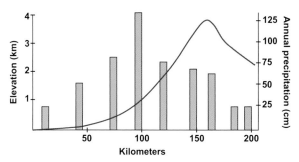

FIGURE 4.8. The rain shadow effect. The solid line represents a cross section of a mountain range and the associated increases in elevation (km) as you go up the mountain. The bars represent the average yearly total rainfall at different elevations in the mountains. Weather (i.e., wind, clouds, and storm systems) in this scheme would be moving in from the left and would be forced up and over the mountains. Note the large increase in precipitation at middle elevations and the decrease in precipitation on the back side (rain shadow side) of the mountain. (Figure by M. P. Marchetti.)

Recall our common-sense physics observations at the beginning of this chapter. Warm, moisture-laden air is blowing across California. It reaches a mountain and is forced to rise up to go over it. As it rises, it expands, cools, and loses its abundant moisture as rain or snow (very similar to the tropical pattern). Now, as it crests the Sierra, it is cool and dry and descends down the back side of the mountains. Yet as it gets lower in elevation, the pressure increases; it warms up and acts like a sponge, literally sucking water out of the surrounding landscape (much like what we saw with the deserts at 30°).

This pattern is called a *rain shadow,* because one side of the mountain receives large amounts of rainfall, whereas the back side (in the shadow) receives very little precipitation (Figure 4.8). The Great Basin Desert owes its existence largely to its geographic position in the "shadow" of the Sierra Nevada. Other rain shadow deserts include Patagonia and the Monte Desert of South America (constituting much of the country of Argentina), and the Gobi Desert, which lies in the rain shadow of the Himalaya Mountains.

NORTH-FACING AND SOUTH-FACING SLOPES

In the northern hemisphere, the sun warms south-facing slopes much more than it does north-facing slopes, resulting in greater temperatures and more rapid drying. A common consequence of this is that very different plant communities characterize the south- and north-facing sides of hills or mountains. This is particularly apparent in relatively dry regions, but it is also visible in temperate and boreal areas as well. The greater moisture availability on north-facing slopes often leads to more lush and dense growth of plants. Animals that require dense vegetation, either for protection from predators or for food, may sometimes be restricted to north-facing slopes.

RAVINES AND RIDGETOPS

The local conditions on various parts of a mountain may also be quite different. Many campers know that cold air settles in the depths of valleys (i.e., cold air sinks), so that if you set your tent at the bottom of a valley or even in a depression in the landscape, you may find it quite cold at night. This is because, when air cools at night, it descends the slopes of a ravine to meet cool, descending air from the other side of the ravine. These cool air masses merge and continue down the valley. If you are camped in the center of a valley, you will not only have cold air at night, but you will also have stronger breezes than if you had camped upslope. While camping in a valley, you may have noticed an evening downslope breeze or a morning upslope breeze. Now you can see why. These types of local microclimatic factors affect the kinds of plants and animals living in an area, as well as the number of campers.

MEDITERRANEAN CLIMATE

The south of France, Italy, California, South Africa, the Chilean Coast, and the southwest coast of Australia all have similar climates and all produce something very important that is dependent on this climate. See if you can figure out what the common thread is among these areas while we describe some of the similarities among their climates. All of these regions have what is called a *Mediterranean climate,* named for the region around the Mediterranean Sea between Europe and Africa. This region is characterized by having a distinct seasonal rainfall pattern, where it rains heavily in the winter and then stops raining altogether during the summer season. Two of the common geographic features that link all the above regions is that they all lie next to an ocean on the west coast of continents (see the area called shrubland in Figure 4.1), and they are all located at about 45° latitude around the globe.

All of these areas also have unique sets of vegetation. In California, this community is called chaparral, in South Africa it is called fynbos, in Chile it is called matorral, in Australia it is called mallee and kwongan. All of these communities are typified by scrubby, dense vegetation that often burn as a means of plant regeneration. The other common feature of these regions, in case you had not guessed it, is that they are, by far, the best regions in the world for producing wine. Wine grapes apparently thrive under the climactic conditions in each of these areas (lucky grapes).

CLIMATE IN CALIFORNIA

Every region of the world has its own special reasons for its climatic peculiarities. Here we present, as an example, the reasons for the climate of a particularly distinctive region of North America, which happens to be where both of

the authors live. A similar account could be written for every region of the world and would help us understand weather phenomena such as hurricanes on the Gulf Coast of the United States, periodic drought in sub-Saharan Africa, and monsoons in Sri Lanka. It is useful to understand your local weather patterns, and to demonstrate, we will use our local weather as an example.

California is hailed internationally as a haven for sun worshippers. We have wonderful weather along much of our southern coast, and the beaches are a haven for bikinis, muscle-bound bodies, and many marine mammals. However, the coast of northern California calls a very different image to our minds—one of fog, lush forests, and lots of rain. Additionally, the climate of California is very seasonal, with relatively dry air and clear skies throughout the summer and fall, and clouds and rain during the winter. A famous Albert Hammond song proclaims "It never rains in California/ But girl don't they warn ya/It pours, man it pours." When it rains in California, it can really pour. But what actually causes the rather substantial change in weather between summer and winter?

To understand this, we need to first briefly understand a couple of additional features of the climate in the North Pacific. At about the latitude of Hawaii, there is a zone of high pressure (a place where air is subsiding, causing the famously nice weather of Hawaii). This area is called the Hawaiian High Pressure Region, or the Hawaiian High. Air that descends here is forced to move either south or north. Air that moves north reaches the Aleutian Islands in Alaska before encountering polar air that is moving south. When these air masses meet, two things happen. First, they have nowhere to go but up, so we find air rising in the region around the Aleutian Islands. This creates a low-pressure region that is referred to as the Aleutian Low, and with it comes associated clouds and storm development. However, because we have warm southern air meeting cold northern air, there is an additional degree of turbulence, so that the storms here also tend to be somewhat big and violent. Aggravating this is the fact that the polar air also is very dry, whereas the air moving north is much more humid and therefore has a very different density; thus, when these air masses meet, they can become quite turbulent. Because the winds here generally move eastward, any storms that form along the Aleutian Low tend to be displaced eastward, and they often drop a lot of rain and snow on the North American continent (i.e., in southern Alaska and British Columbia).

Now, this all gets interesting (and particularly relevant) when we superimpose the seasonal shifting of the Earth relative to the sun onto the position of the Aleutian Low and the related weather patterns. You may remember that the position of the sun relative to Earth shifts seasonally, from as far north as the Tropic of Cancer to as far south as the Tropic of Capricorn. As the sun shifts north and south, the major bands of rising and subsiding air masses also shift north and south (although not as much as the sun does). However, the major storm generator for the Pacific Northwest, the Aleutian low-pressure region, shifts from about 60° N latitude in the summer to about 50° N latitude in the winter. So, in summer, this low pressure region lies sufficiently north so that the storms generated there collide with North America in Alaska and Canada. However, winter brings a double whammy to this scenario.

First, the air that is descending from the polar regions is colder in winter than in summer. When this frigidly cold air meets air moving north from the Hawaiian High, the great difference in their temperatures results in stronger storms than in summer. Additionally, the low-pressure zone itself has shifted south in response to the sun's southerly migration, and the storms that are generated then plow right into the north coastal regions of our fair country. This is why northern California and Oregon get so much rain in the winter while staying warm and dry in the summer. Fortunately for people in southern California, the storm tracks generally reach no further south than the middle of the state. Of course, these storms are responsible for the multimillion dollar skiing industry around Lake Tahoe as well.

BIOLOGICAL CONSEQUENCES OF GLOBAL CLIMATE PATTERNS

So far, we have learned much about the global distribution of climate in this chapter. What can this tell us about the distribution of plant and animal life? Because climate strongly influences things like soil formation, we might expect to find very different soils under different climatic regimes. This indeed is the case, and so it is not too surprising that climate also strongly influences the distribution of plants. For example, tropical trees and vines are unlikely to survive in a desert environment. Similarly, many desert plant species have very poor defenses against plant diseases that thrive in moisture, and so desert plants generally don't survive well in wet climates.

Understanding climate therefore gives us vital information about where you would find animals that are particularly good at surviving on little water or animals that require high humidity and/or frequent rains. If you are studying animals that live at high latitudes, understanding climatic conditions for that region helps you understand much of their basic biology. If you study marine systems, oceanic patterns provide important insights into the movement of fish and plankton (consider the global circulation of water) or their food items (think upwelling).

CONVERGENT EVOLUTION

As strange as it might seem, we are now going to make a connection between what we have learned about evolution in previous chapters and what we have learned about global climate. It is important to realize that the similar climates found at distant parts of the globe often lead to very similar habitats (i.e., tropical rainforests in South America, Africa, Southeast Asia, and Australia). Within these similar habitats, it is not uncommon to find that the plants and animals have evolved similar strategies for interacting with their surroundings. When organisms evolve similar means of coping with their environments in different parts of the world, we call the process *convergent evolution*.

A great example of convergent evolution can be seen by comparing mammals found in Australia with other mammals from around the world. Before we do this, it is useful to realize that there are actually two very different ways of being a mammal. In the majority of the world's mammals, a developing embryo is connected to the mother through the placenta inside the womb. But there is a small group of mammals called the marsupials, where there is no placental connection. Instead, at a very young age, the embryo actually crawls out of the mother's womb and climbs up into her pouch, where the embryo grows and finishes development by nursing at the mother's teat.

Australia, as you might remember, is a large continental island that broke off from Asia many millions of years ago. Interestingly, at the time Australia broke from Asia, the only mammals that had made it to Australia were marsupials. In the rest of the world, over many millions of years, placental mammals proved to be better competitors and eventually replaced most marsupial mammals, but not in Australia. Instead, the vast majority of Australia's native mammals are marsupials.

Now the interesting part about convergent evolution can be seen by recognizing that Australia's climate produced similar biomes to other parts of the world (e.g., deserts, rainforests, Mediterranean scrub). We expect that animals living in Australia's deserts face the same challenges as other desert animals around the world and might therefore have come to have similar characteristics by the process of natural selection.

We can see evidence of this by comparing marsupial mammals native to Australia with other mammals. For example, the grey wolf (carnivorous, medium-sized predatory placental mammal found in North America and Eurasia) is similar to the Tasmanian wolf (carnivorous, medium-sized predatory marsupial mammal found in Australia), whereas flying squirrels (placental mammal that glides between trees found in North America) are similar to flying phalangers (squirrel-like marsupial mammal that glides between trees in Australia). Numerous other examples of this type of convergent evolution exist for birds, reptiles, fishes, plants, and other groups.

It is also important to remember that, as a result of the process of evolution through natural selection, organisms become increasingly well adapted to their environments over time. The large variations in local climate that we have discussed above provide an endless array of environments to which species may adapt. Thus, we find different lizard species in Baja Mexico than we do in Nevada or Utah or Florida. We see different plants growing in coastal, prairie, and mountainous areas within a single state, and sometimes even on different faces of a single hillside. Understanding local and regional climate and the factors governing these can therefore provide insight into the causes of biodiversity we see around the globe.

GLOBAL CLIMATE CHANGE

The above description of how the global climate regime works makes it sound as if we really understand how climate works. In a broad sense we do: the basics of climate are determined by the physics of water, the angle and rotation of the Earth, the location of landforms, and other "fixed" entities. However, when you realize that the local manifestation of climate is weather, then you understand that climate is only predictable on a fairly broad scale (i.e., weather predictions in your area are only really good for a day or two ahead of time, with prediction success getting weaker as time increases). But that predictability is enough to allow us to determine where most animals and plants are likely to be found: polar bears in the Arctic, redwood trees on the California coast, and parrots in the tropics. It is also understandable enough that we humans have figured out how to live in most terrestrial areas by adapting either ourselves or our lifestyles to the local climate. Unfortunately, our nice, somewhat predictable global climate system is becoming less predictable, through the process of climate change, also known as global warming. Actually, a better term for what is happening may be "global weirding," which was coined by Hunter Lovins, cofounder of the Rocky Mountain Institute, and popularized by political columnist Thomas Friedman.

According to Lovins and Friedman, global warming is much too mild and innocuous a term for what is happening to the planet. Global weather patterns are becoming more extreme, more violent, less predictable, and therefore "weirder." At the same time, ocean currents are becoming less dependable, and even the solid land upon which we live is becoming less reliable as a safe home. As a result of these changes in climate, flowers are blooming earlier in the year in many places, temperate bird and butterfly species are being seen further north, malaria and other diseases are spreading to new environments, and many plant and animal species are facing extinction as their habitats shrink. More importantly to us as humans, droughts are becoming more frequent and more severe, hurricanes are bigger and more destructive, the Arctic ice cap is melting, as are most glaciers, and to top it all off, the global sea level is rising.

The basic cause of these global changes is an accelerating increase in carbon dioxide and other gases in the atmosphere, largely caused by the burning of fossil fuels (e.g., coal, oil, gasoline, etc.). The concentration of carbon dioxide in the atmosphere has risen about 30 percent since the late 1800s, and it is already higher than it has been for at least 650,000 years. And it is still rising very rapidly.

In the past, the carbon dioxide level in the atmosphere was regulated by plants, especially algae in the ocean, which use it in the process of photosynthesis. Some of the newly created carbon-rich plant tissues wind up falling into places where they cannot be recycled: low-oxygen areas of the deep ocean, peat bogs, and similar areas. This locked up carbon dioxide is essentially stored in the Earth's crust and is eventually converted after millions of years into substances such as oil, natural gas, and coal. When we burn

the oil, we are essentially liberating huge quantities of ancient carbon dioxide that have lain dormant under the Earth's surface for centuries. This "new" liberation of "old" carbon is so large that it is way beyond the capacity of current plant life to absorb. This excess gas is a problem because carbon dioxide acts as a "greenhouse gas" and traps heat from the sun in the atmosphere—heat that would normally escape into space. As a result, the entire planet gets warmer, 10–12°F on average over the next century.

Thanks to the global climate system described in the pages above, however, the Earth does not warm up evenly or feel the effects of the temperature increases uniformly because winds blow and ocean currents move large masses of air and water long distances. Thus, increased temperatures in the tropics are likely to increase the frequency and intensity of typhoons (caused by the interaction of atmosphere and ocean), even if the basic wind circulation patterns don't change. An increase in temperature in the Arctic causes the ice cap and snow to melt and may cause less rain to fall in the winter in Washington and Oregon.

This causes further problems because these features (snow and ice) help keep the planet cooler by reflecting solar heat back out into space. Instead, this heat is now being absorbed by the dark land and water that was once covered by ice. One consequence of this is the melting of the permafrost, a vast reservoir of ice and ancient carbon (stored as dead plant material or peat) in the tundra. Melting permafrost is likely to have two consequences. The first is the release of yet more carbon dioxide into the atmosphere, and the second is the release of large quantities of fresh water into the ocean. The latter has the potential to physically alter the way ocean currents flow because they are driven by wind and water density (warm and fresh water are less dense than cold and salt water).

As you can see, this chain of events provides a good indication of why the process of climate change is actually getting faster over time and the local weather is becoming more variable and more severe. As Hurricane Katrina demonstrated to New Orleans and the rest of the world, global climate change is not an abstract concept cooked up by nerdy academic scientists—it is real, and it is happening before our very eyes.

Hurricane Katrina also demonstrated another, more subtle phenomenon: sea level is rising as the world's ice melts. It is predicted that sea level will rise by a meter or more in the coming century. This may not seem like much until you realize that much of the world's human population lives below, at, or just barely above present sea level. Low-lying countries such as Bangladesh are increasingly subject to storm surges and high tides, which can flood huge areas where people live and work. Coastal cities are built where they are because of easy access to the ocean; many of them are now threatened by flooding from sea level rise. New York City, for example, will not only have low-lying areas flood with rising sea levels, but the entire underground subway system will also be increasingly difficult to maintain free of water.

If global climate change sounds to you like a disaster of epic proportions for humans and our way of life, it is even more so for the natural world, which is much less adaptable. We have confined much of the Earth's biodiversity into relatively small parks and natural areas. Although protected, these areas tend to be small and distant from other parks or natural areas. As a moist forest dries out or a freshwater spring stops flowing, where are the endemic animals and plants going to go? Extinction is the unfortunate but likely outcome for many organisms. Zoos, aquariums, and botanical gardens are ex-situ (out of place) conservation options for the lucky few that we manage to scoop up before

they disappear, but the reality of captivity may not be much better than extinction. The polar bear, for example, without Arctic ice, is likely to go extinct in the wild; this basic fact has caused it to be listed in the United States as a threatened species, and the bear is now a poster child for global climate change.

The enormous and pervasive effects of climate change require global action, especially to reduce carbon dioxide and other "greenhouse" gas emissions. The continued failure of the biggest carbon emitting countries, such as the United States and China, to face up in a major fashion to the global climate realities that have been explained by multitudes of scientists suggests that changing our ways is not going to be easy, or even likely, until nature gives us no choice.

FURTHER READING

Crawford, W. P. 1992. Mariner's weather. W.W. Norton and Co. New York. *This book for sailors has a wonderful explanation of common (and uncommon) weather events written for people with little or no science background.*

Dow, K, and T. E. Dowling. 2007. The atlas of global climate change. University of California Press. Berkeley. *With the use of dramatic graphics, climate change and its potential effects are explained in an understandable fashion.*

Forthergill, A., V. Berlowitz, M. Brownlow, H. Cordey, J. Keeling, and M. Linfield. 2006. Planet Earth. University of California Press. Berkeley. *A celebration of our planet's natural systems, with marvelous images and straightforward text; this book would be a good one to read as a supplement to this text.*

Pianka, E. R. 1988. Evolutionary ecology, fourth edition. Harper and Row Publishing. New York. *This may be a little more than some people want to tackle, but this book gives a concise yet understandable description of global climate if you want to learn more.*

5

Ecology

So far we have looked at the process of evolution through natural selection (Chapter 2), how species develop (Chapter 3), and how climate can set the stage for the plants and animals that live in a particular place (Chapter 4). This chapter and the next provide some basic tools and language for understanding the processes by which species adapt to their environment and interact with members of their own and other species. That is to say, these chapters describe the study of ecology. A solid grounding in the fundamentals of ecology is essential for understanding the modern science of conservation biology.

In a general sense the study of ecology can be categorized by four levels of organization:

Organismal (individual) ecology: how individual organisms interact with their local environment.

Population ecology: how populations grow and vary over time.

Community ecology: how different species interact with each other and divide resources.

Ecosystem/landscape ecology: how energy and nutrients move through ecosystems.

We will tackle organismal and population ecology in this chapter and the other two in the next chapter. By the end of these two chapters you will see that although we can frame ecological ideas in these four categories, many studies (particularly in conservation biology) cross these lines and combine multiple ecological levels.

ORGANISMAL ECOLOGY

Organismal ecology is *the study of how individual organisms adapt and thrive in their particular environment*. In other words, it describes how individuals make a home for themselves in the place they live. One way to approach this concept is to think about the ways the environment constrains how an organism lives. An intuitive example comes from thinking about a camping expedition. Let's say you are taking a summer wilderness-survival course in Death Valley, California, one of hottest and driest deserts of the world. For the last week of the course you will go on a solo hiking trip, but you are allowed to take only five things with you. What would you take? Typical answers include such items as water, food, shade, sunscreen, sun hat, light-colored and lightweight clothing, and more water. How would your list change if the course instead was being taught at 10,000 feet in the Rocky Mountains? Typical answers would include food, water purifier, sleeping bag, tent, warm clothing, and insect repellent. How would it change if instead you were going trekking in the rainforests of Costa Rica? There you would include food, rain gear, lightweight clothing, water purifier, waterproof boots, insect repellent, and a rainproof tent. How might the list of necessary desert gear change if you were from a native culture endemic to that section of desert? How would desert gear change if you were taking your 90-year-old grandmother with you? Note that these lists keep changing, for many reasons. Sometimes they change because of the physiological and behavioral constraints the environment places on organisms living there, and sometimes they change because of inherent constraints of individuals. We are going to examine some of these constraints and how they relate to an individual organism's ecology.

DESERT LIVING

One of the easier places to examine aspects of organismal ecology is in the desert. Deserts are harsh environments, with three main environmental constraints to living there. The first is, of course, lack of water. Desert environments are actually defined by having annual rainfall totals below a certain value (often less than 254 mm, or 10 inches, per year). So living in a desert means you have to deal with little available water as a fact of life. But this does not mean that it never rains in the desert. In fact, sometimes it rains huge amounts in very short periods. This can result in very quick runoff and flash floods. If you have ever visited a desert of any kind, you may have seen the visible landscape marks of these floods: dry water courses called arroyos or desert washes.

A second constraint to living in the desert is temperature. Deserts are often very hot. In fact the hottest temperature recorded on the planet is from the desert region in Libya: above 135°F (47°C). Yet heat is not the only temperature

issue in deserts; they typically get bitterly cold at night because of rapid heat loss from lack of water and vegetation. Both the heat and cold are exacerbated by the fact that, because there is little shade or cover in the desert, larger organisms, such as humans, are constantly exposed to the elements.

A final constraint to living in the desert is the fact that too much sunlight can be a bad thing. For you and me, excess sun can cause our skin to burn and blister, which can eventually lead to skin cancer and death. Too much sun is a problem for plants as well as animals. Not only can their tissue be damaged directly by excess solar radiation, but they can also lose excessive amounts of water through photosynthesis. Sunlight turns on the mechanism of photosynthesis, which requires water, so desert plants are repeatedly facing a crisis: make food and lose water, or don't make food and save water. To top it all off, deserts are generally sunny year round, so there is not much seasonal relief from the sun's rays. Too much sunlight poses problems for the organisms that live in the desert.

KANGAROO RATS

A good example of a desert animal that deals successfully with these environmental constraints is Mirriam's Kangaroo rat (*Dipodomys mirriami*) (Figure 5.1). Interestingly, the Latin root-words for the name of this genus of rodents give us some information about the critter. *Dipo* means thirsty and *mys* means mouse, and the name "thirsty mouse" seems reasonable for mice that live in a desert. Yet these cute little rodents rarely go thirsty; they are marvels of evolution in their adaptations for conserving water.

To see this, we first need to examine water loss in mammals. Mammals lose water in one of three ways: they breathe it out, they sweat it out, and they pee it out. Water is lost through breathing, which mammals do to acquire the oxygen necessary for cellular metabolism and to get rid of carbon dioxide, a waste product. Water is lost in the process because it is a physiological requirement that oxygen be absorbed

FIGURE 5.1. Stephens Kangaroo rat *(Dipodomys stephensi)* pictured here from San Diego County, California, is a very close relative of Mirriam's Kangaroo rat, discussed in the text. (Photo by Cheryl Brehme; used with permission.)

across a wet membrane. The insides of mammalian lungs have a nice wet membrane, and when we breathe out, some water is exhaled as water vapor. Mammals generally sweat to cool their bodies down because evaporation of water removes heat. Animals urinate to remove nitrogenous waste from their bodies because nitrogen is a toxic byproduct of digesting food, particularly proteins. If a mammal did not urinate (or if its kidneys shut down), it would quickly build up too much nitrogen, its blood would become toxic, and it would die.

If we humans get stuck in a desert and run out of water, we can only do a few things to help limit our water loss, mostly through behavior. Mainly, we can find some shade and remain as inactive as possible to reduce sweating and breathing. In terms of water loss through urine, we do not have many options available to us, because we have no way of reabsorbing water from our bladder. In dire straits, we can drink our own urine, but that is not a recommended or particularly palatable solution.

If we look to the kangaroo rat, however, we can see that evolution has equipped this animal with a host of amazing adaptations to help it conserve precious water in its desert environment. First we turn to the kangaroo rat's nose, through which it breathes. The rat's nasal passages are set up in such a way that it is able to conserve and recover some of the water from its

breath. Its nose works as a countercurrent heat exchanger (similar to a car's radiator) in that the air entering and leaving the rat's lungs has to follow a convoluted pathway. Along the passage a heat gradient is set up, with breath getting cooler as it gets closer to the outside. This heat gradient allows time for the water vapor to condense inside the nasal passage (remember from Chapter 4 that cool air holds less water than warm air). This relatively simple mechanism allows the kangaroo rat to recover somewhere between 50 percent and 80 percent of the water in its breath, making it very water-efficient.

Another astonishing fact about this kangaroo rat is that it does not drink water. In the deserts where these rodents live, they can't rely on free water because they rarely encounter it. If they don't drink water, why don't they die of thirst, you might ask. One way they get water is through their food. When they break down food during digestion, one of the byproducts is water. For demonstration purposes, we will represent all food using the simple sugar glucose, which has the chemical formula $C_6H_{12}O_6$. Digestion of glucose requires oxygen (O_2) and breaks this sugar into carbon dioxide (CO_2) and water (H_2O), releasing energy. If you do the math, you can see that, for every molecule of glucose digested, we get six water molecules. This process is not unique to kangaroo rats; it happens when we digest food as well, but these rats are more efficient at generating and using this water. In addition, kangaroo rats are seed eaters and they will often bury, or "cache," their seeds underground, where the temperature is significantly cooler. These seeds are bone-dry when collected at the desert surface, but when they are put into cool underground caches, they act like little sponges, sucking up some of the moisture from the surrounding soil. So kangaroo rats also get some water from the way they handle and process food.

When we look at water loss through sweating, we find that kangaroo rats have solved this problem by staying in their burrows, limiting the time they spend out in the hot sun. In fact, they don't sweat much because they run around at night, when it is significantly cooler and there is no sunlight to dry them out. Finally, in terms of water loss through urine, the kangaroo rats have a specialized kidney (with an elongated loop of Henle) that is extremely efficient at removing the water from their nitrogenous waste (five times more efficient than that of humans). So efficient are the kidneys that the kangaroo rat's urine is excreted more as a paste than as a liquid, saving it large volumes of water. With all of these physiological and behavioral adaptations, the kangaroo rat is a marvel of water conservation, which allows it to live in the harsh, dry desert.

FIGURE 5.2. Desert iguana *(Dipsosaurus dorsalis)*. (Photo by Jackson Shedd; used with permission.)

DESERT IGUANA

For a second example of environmental adaptation we turn again to the desert, this time to examine the physiological ecology of the desert iguana *(Dipsosaurus dorsalis)* (Figure 5.2). You may recall that in Latin *dipo* means thirsty, and you may also know that *saurus* means lizard (as in dino–saurs, or the terrible–lizards), so this genus consists of thirsty lizards. *D. dorsalis* lives in the deserts of Arizona, southern California, and northern Mexico. This lizard is an herbivore, eating only tough desert plants. On its own, however, the desert iguana cannot digest cellulose, which is the main component of plant matter. This would be a problem except

that, like many herbivores, the lizard has special bacteria in its gut that have the ability to digest cellulose. These bacteria work most efficiently at around 42°C (98°F). Curiously, the lizards are *ectotherms*, meaning that their body temperature depends on the external environment, and most of the time the air temperature where they live is either hotter or colder than 42°C. (Humans and other mammals, in contrast, are *endotherms*, regulating their body temperature to a constant level regardless of the external environment.) But if we look at a graph (see Figure 5.3) showing desert surface temperature, lizard body temperature, and the lethal body temperature for the lizard, we see an interesting pattern. Even

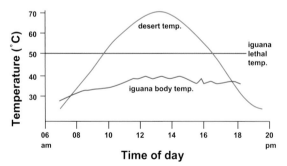

FIGURE 5.3. A graph showing the relative stability of the internal body temperature (°C) of a desert iguana (green line) throughout a typical midsummer day compared with the external desert temperature (red line). The blue line represents the upper lethal body temperature limit for the desert iguana. (Modified after figure from Porter, W. P., J. W. Mitchell, W. A. Beckman, and C. B. DeWitt. 1973. Behavioral implications of mechanistic ecology: thermal and behavioral modeling of desert ectotherms and their microenvironment. Oecologia 13:1–54.)

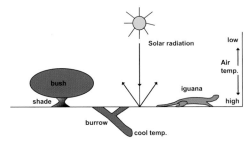

FIGURE 5.4. Behavioral options for the desert iguana *(Dipsosaurus dorsalis)* to thermoregulate. Note that the highest temperatures in the desert are often found directly at the surface of the soil.

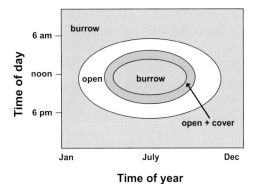

FIGURE 5.5. Map of yearly and daily behavior for the desert iguana *(Dipsosaurus dorsalis)*. Note that during the heat of the summer, the iguanas spend only a minimal amount of time in the open. (Modified after figure from Porter, W. P., J. W. Mitchell, W. A. Beckman, and C. B. DeWitt. 1973. Behavioral implications of mechanistic ecology: thermal and behavioral modeling of desert ectotherms and their microenvironment. Oecologia 13:1–54.)

though the desert gets very hot (well over the lethal temperature for the lizard), the body temperature of the lizard stays near 42°C for most of the day, meaning that the lizard's internal bacterial colony is working near its optimum.

How does the ectotherm lizard do this? The answer lies in its ability to regulate its body temperature by a variety of behavioral strategies. Figure 5.4 shows some of the options available. It turns out that the hottest place in the desert is at the surface (i.e., right on top of the sand), but if you can raise yourself up even a little bit, you can be slightly cooler. The lizard takes advantage of this by doing pushups to lift its body off the sand. In addition, the lizard can change its orientation to the sun, much like you might see people do at the beach while tanning themselves. Of course, there are also low shrubs that provide some shade as well as underground burrows where the lizards can stay cool when it is really hot. If we were to make a behavioral "map" of what the desert iguana does each day, it would look something like Figure 5.5. This "map" shows where the lizard is located throughout each day over a year's time. We can see that in January the iguana is spending all day in its burrow, whereas in July it comes out into the sunlight in the morning, spends a few hours under shade, then retreats to its burrow during the hottest part of the day, only to re-emerge late in the afternoon and early evening. By following this behavioral "map," the iguana is able to maintain its internal body temperature very close to the optimum for maximum bacterial

digestive efficiency and live happily under the variable temperature regime of the desert.

THE NICHE

So far in this chapter you have seen that individual species have a suite of physical and behavioral adaptations that allow them to fit particularly well with their environment. Each species has its own unique suite of these adaptations. Ecologists have long noted this and have coined a term in an attempt to describe this phenomenon: *the ecological niche*. This term is a particularly tricky one because it is used to describe a multifaceted set of traits unique to each individual species. Some people like to think of the niche as the sum total of the ways a species uses resources, and others think of it as the pattern of how a species lives in its community. A more formal definition is *the activities and relationships of an organism constrained by physical and biological processes*. It is sometimes easiest to think of the niche in terms of answers to a series of questions. First, where would you find the organism (i.e., what is its environmental address)? For example, if you wanted to find a salamander, where would you look for it? Would you look for it sunning itself on hot rocks, or would you look for it in wet places, such as under a rotten log? The answer depends somewhat on the species of salamander, but in general the best places to look would be in damp or wet environments and not out in the open. Many people know that salamanders are moist-skinned and would dry out if exposed to the sun for very long; that is, we know the general "address" for amphibians. Now if we combine this knowledge with answers to the following questions, we start to get a handle on a species' niche.

1. What does it eat, when does it eat, and how much does it eat?

2. When does it move and how far does it move?

3. How, when, and where does it reproduce?

4. Who eats it and how it does it avoid being eaten?

5. What other species does it interact with and how, when, and why does it interact with them?

The niche has been defined in various ways over the years, and at one point was defined by the famous ecologist G. E. Hutchinson as "an n-dimensional hyper-volume." Because most ordinary humans have a hard time picturing things in more than three or four dimensions (the fourth being time, of course), this definition is hard to comprehend. But if we look at an example using only three dimensions, perhaps we can get a feel for what Dr. Hutchinson was getting at. Our example for this involves a species that many people have some experience with, especially if they have ever been fishing, namely the rainbow trout (*Oncorhynchus mykiss*). The rainbow trout is a widely introduced sport fish that inhabits streams and rivers. One way to start to describe this fish's niche is to go to a stream that has trout in·it and dive into the water with a mask and snorkel. Then, every time you see a rainbow trout, record three pieces of information: water temperature, speed of the stream (water velocity), and the amount of shade above the

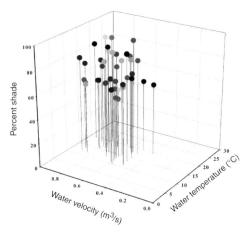

FIGURE 5.6. Hypothetical rainbow trout (*Oncorhynchus mykiss*) niche graph based on three environmental variables: water velocity, percentage of shaded cover, and water temperature. Note that each dot represents a measurement from an individual trout; the drop-down lines from the dots are present to help interpret the figure. (Figure by M. P. Marchetti.)

stream. If you did this for 50 different trout, you would have data that we can graph in three dimensions. The graph from this kind of field experiment might look like Figure 5.6. In this graph each of the dots represents the three pieces of data from one trout. We can see that the points tend to cluster in certain areas of the graph. The trout seem to like cold water that moves fast and has lots of shade over it. This is a three-dimensional niche graph for these rainbow trout. Perhaps we can imagine if we did this for more than three variables that we could get something like Figure 5.6 but much more complicated. This is essentially what Hutchinson meant by an "n-dimensional hyper-volume." He meant a combination of all the physical and environmental variables considered together.

Another layer of difficulty with the niche concept is that what we have described so far has been called the *fundamental niche*. The fundamental niche considers what the species *could* do, or the range of possibilities its biology allows, but not necessarily what it does in reality. For example the fundamental niche of rainbow trout does not include low-elevation rivers where the water temperature gets over their physiological limit (i.e., approximately 21°C). Yet the fundamental niche of these trout could include a warm river during the winter months, when the water is cold, but trout are rarely found there because the water gets too warm for them during the summer months. This is in contrast to the *realized niche,* which is defined in terms of what the species actually does when it is constrained by other creatures.

An example should help clear this up. If we think about the mountain lion *(Felis concolor)* in the western United States, we find that these cats can live in most areas of the region: including mountains, valleys, coastal areas, and even parts of the desert. This (combined with all the other variables) describes part of its fundamental niche. In contrast, the realized niche of the species would be where we actually find it when we go looking for it. For example, we are unlikely to find many mountain lions in downtown San Francisco, even though they could potentially live there and probably did sometime in the past. Their niche is constrained, in this case by human beings, to a much smaller subset of where they could exist. Likewise, rainbow trout may be absent from some perfectly suitable cold water streams because of the presence of brook trout *(Salvelinus fontinalis),* which can exclude them through competition and predation.

HABITAT AND RANGE

In our definition of niche so far, we have assiduously avoided using two related yet different terms: *habitat* and *range.* How is a habitat different from a niche? And how does each of these differ from a species' range? The way we defined niche above was pretty abstract, in that it brought together all of a species' requirements. The niche is not something that you can see. You cannot go to a niche, nor is it concrete in space or time—and it may be altered with evolutionary changes in the species. It is more a characteristic of the organism than of the environment.

FIGURE 5.7. A portion of a bighorn sheep *(Ovis canadensis)* herd in Anza Borrego Desert State Park, California. If you look carefully, you should be able to see six animals in the photograph. (Photo by M. P. Marchetti.)

This is in contrast with the term *habitat.* A habitat is an actual place; you can see it and you can go to it. Habitats are typically named according to either the dominant vegetation or life-form (e.g., coniferous forest, coral reef, mangrove forest, desert sage-scrub, kelp forest, tall-grass prairie) or in relationship to physical locations or traits (e.g., alpine stream, near-shore sandy-bottom marine, freshwater spring, desert floor, alkali lake). Habitats are similar to biomes (see Chapter 3) but are generally thought of on a much smaller scale. Biomes are large global regions, whereas habitats are more local in scale. A species that lives in a particular habitat must find all the components of its realized niche in that habitat.

Range, on the other hand, is a term that describes the map location where you can find a species. This can even include multiple habitats. If you place a dot on a map for every location where you find a particular species and then draw a line around all of them, you essentially get a range for that species. A few visual examples should help us to sort this all out.

The first example is the desert bighorn sheep *(Ovis canadensis),* a large grazing mammal famous for its big curled horns (Figure 5.7). Its niche is that of a wide-ranging herbivore adapted for living in mountains surrounded by desert

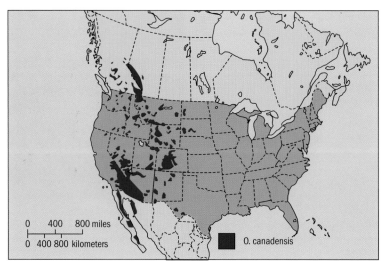

FIGURE 5.8. Map of bighorn sheep *(Ovis canadensis)* range in North America. (Reprinted with permission from Chapman, J. A., and G. A. Feldhamer, eds. 1982. Wild mammals of North America: biology management economics. Johns Hopkins University Press. Baltimore.)

(its habitat). Among its niche requirements are free-standing water to drink (e.g., water holes) and rocky ledges to which the animals can leap to escape predators. It also appears to be more physiologically tolerant of heat and dry conditions than other kinds of bighorn sheep. Historically, adults dispersed by moving across desert regions in the spring, when water and food were plentiful, but they still spent most of their time in the mountains, where they could more easily avoid predators (including humans). The range of the desert bighorn encompasses the dry mountains of the western portion of North America (Figure 5.8).

Our second example is the desert iguana, introduced previously in this chapter. The niche description for this lizard is an ectotherm that munches on succulent desert plants and has behavioral adaptations to cope with extreme temperatures and to avoid predators such as hawks and coyotes. Its habitat is creosote brush desert with hummocks of loose sand and patches of firm ground with scattered rocks. Its range is limited to the Mojave and Sonoran deserts of the southwestern United States and northwestern Mexico (Figure 5.9).

For our final example we turn to a rare case, the Devils Hole pupfish *(Cyprinodon diabolis)*, a species whose niche, habitat, and range all essentially converge on one another. The Devils Hole pupfish has the narrowest habitat and range of any known vertebrate species. You can find this handsome half-inch-long critter in only one natural location, which is a place that seems unlikely to contain any fish at all. To get there you need to drive an hour or so northwest of Las Vegas, Nevada, out into the middle of a dry desert valley. If you look very carefully at the edge of a hillside, you might see a fenced-off area the size of a baseball infield. If you are able to gain access through this chain-link fence and look down into a crack in the Earth, you will see a tiny pool, roughly the size of a double bed, which composes the worldwide niche, habitat, and range of the Devils Hole pupfish (Figure 5.10). The entire species is contained in this area. The pool is the top of a submerged cave, so the water goes down many meters, but the fish is largely confined to a shallow shelf where the algae that it eats can grow. The niche of the species is also quite limited: it is an herbivore specialized for living in an environment in which temperatures

FIGURE 5.9. Approximate range of the desert iguana (*Dipsosaurus dorsalis*). (Figure by M. P. Marchetti.)

FIGURE 5.10. The entire niche, habitat, and range of the Devils Hole pupfish (*Cyprinodon diabolis*). Note that the entire area of this spring is only 2 meters by 4 meters. (Photo by Sean Lema; used with permission.)

are constant and food is scarce. The alga it eats grows as the result of the sunlight that reaches the shelf for a few hours each day. The range of this pupfish is clearly just a dot on a map. This fish has been a focus of conservation concern for almost 40 years, because of the fact that groundwater pumping in the early 1970s lowered the water level, exposing its shelf. The pupfish was pulled from the brink of extinction by a decision of the U.S. Supreme Court, which stopped the groundwater pumping; eventually the water levels rose and the population recovered. Yet the future of this fascinating vertebrate species remains uncertain because pumping of groundwater for use in Las Vegas continues in the surrounding region.

The examples of mountain lions, desert bighorn sheep, and Devils Hole pupfish all provide a good generalization about conservation based on ecological information: the smaller the niche, habitat, and range of a species, the more vulnerable it is to extirpation by human activities.

POPULATION ECOLOGY

So far we have seen how individual organisms interact with their environment to survive and reproduce. Now we want to move to the next level up the ecological hierarchy and examine the ecology of populations. A good ecological definition of a population is *a group of individuals of one species that live and reproduce together*. Populations are important to ecology and conservation for a number of reasons. Like species, they are a basic unit of biology, and as such populations have the potential to exert a large influence on the environment. In addition, understanding the ecology of populations helps us understand higher levels of ecological organization (communities and ecosystems). Many important conservation issues are contained in the study of populations, as we shall see.

One of the first things to realize about populations is that, regardless of the species involved, a population has a set of characteristics that scientists can measure. There are three easy characteristics that we focus on initially. The first is perhaps the most obvious one, *population size*, or the number of individuals in the group. If you think about it for a minute, the number of individuals is a pretty important value for conservation work. If a population is very large, it can overrun and potentially alter its environment and therefore become a conservation concern. What would happen if the seas were overrun with squid, or your backyard overrun by rabbits, or like a 1950s disaster movie, a town entirely consumed by spiders? On the opposite side of the coin, if a population is extremely small, then it may be more likely to go extinct and thus become a conservation issue. A large fraction of conservation research is directed toward the ecology of small populations, as we will see in coming chapters. We will also see, later in this chapter, that keeping track of the number of individuals through time can be a bit of work that will require some simple mathematics.

A second population characteristic to keep in mind is the population's *sex ratio*. This is something you could go out and measure by just counting the number of males and the number of females. Clearly, if you have a population with a large number of females and few males, it is likely to grow more quickly than if you had a population dominated by many males and with few females. The sex ratio is important for populations of sexually reproducing organisms but is not a concern for asexual organisms.

The third simple characteristic of populations is *dispersion,* or how the individuals are arranged across the landscape. There are three general dispersion patterns of populations. The first, called random or haphazard dispersion, is often found where individuals in the population do not have strong interactions with each other. This pattern is relatively uncommon in the natural world. The second, called uniform or even dispersion, is generally the result of strong interactions among individuals. This type of dispersion is common among plants, attached marine invertebrates, and highly territorial organisms. The third pattern is called clumped or clustered dispersion, and it is often due to resources that are clumped or clustered across a landscape. This pattern is very common in nature (trees in a forest, whales in a pod, birds in a flock, insect swarms, fish schools, elephant herds, etc.).

POPULATION DYNAMICS

We have seen that populations can have different characteristics, but we are going to focus the remainder of this chapter on population size, the most important characteristic for conservation. The study of how populations grow and change over time is called *population dynamics*. Population studies play a very large role in our lives even though we may not realize it. Studies of human population dynamics (demography) are largely responsible for the way health, auto, and life insurance companies decide their rates. For example, health insurance demographers studied the U.S. population in great detail and found out that smokers are much more likely than nonsmokers to develop serious health-related problems (diabetes, cancer, high blood pressure, etc.) later in life. As a result, the insurance companies now charge smokers more money to cover their future bills. Similarly, car insurance companies studied people's driving records and realized that 16–22-year-old males are most likely to cause accidents and thus charge them more money for car insurance.

Population studies examine a whole suite of factors but we are only going to look at age structure and survivorship. The first is the *age structure* of the population, or the number of individuals in each age category (Figure 5.11). This is a very important piece of information because not all individuals in a population are equal: they often have vast differences in some of their vital rates. There are two *vital rates* we want to examine; the first is the fecundity rate. Fecundity is essentially the birth rate, or the average number of offspring per individual per year. Not all individuals in a population are able to have offspring; some are pre-reproductive, or sexually immature, and others are post-reproductive, or past the age where they can bear offspring. Notice also that the fecundity rate may differ between males and females or between age-20 females and age-30 females. The left side of Figure 5.11 shows an age pyramid for a human population that is likely to grow rapidly. What parts of it suggest rapid growth? Rapid population growth is likely when there many young individuals that will soon enter the reproductive ages (i.e., a delta- or triangle-shaped age structure) and few individuals that

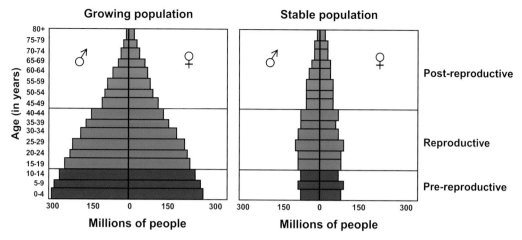

FIGURE 5.11. Hypothetical age pyramids for both a growing and a stable population. The left-hand side of each Figure represents males in the population and the right-hand side females. The length of each bar represents the total number of individuals in each age class. The bars are colored based on reproductive status. (Figure by M. P. Marchetti.)

are likely to die (i.e., older individuals). Compare that with the age structure in the right side of Figure 5.11, which is characteristic of a population with stable numbers through time. Notice the differences. A second vital rate is the mortality rate, or death rate, which is defined as the number of individuals in each age category dying every year. These rates too will vary by sex and by age, creating an even more dynamic snapshot of a population.

The second important population feature, called *survivorship,* is directly related to mortality (death) rates. Survivorship curves show us how the mortality rate of a population changes with the age of the individuals. For example, we know that in humans the mortality rate becomes pretty high (near 100%) when we get older than 95 and that young children in Western countries do not die very often. But is that pattern the same for all organisms in nature? The answer is a resounding no. There are essentially three different types of survivorship patterns, which have been given the amazingly creative names of Types I, II, and III. Figure 5.12 shows all three curves. A Type I survivorship curve is indicative of organisms that have a low mortality rate until adulthood and

then after a certain age survival decreases rapidly. Humans tend to fit this type of curve, as do many large mammals. The Type II curve shows a constant mortality rate regardless of age. Which organisms fit this type? The answer is that very few plants or animals have this type of survivorship. One example that seems to fit is the population of sea anemones living in a rocky intertidal habitat. The anemone's major source of mortality is being crushed and pulverized by pounding waves or logs, and the chance that this will happen is essentially constant throughout the anemone's life. The Type III curve shows very high mortality early in life, but once the organism reaches a certain age, it is likely to live a long time (or at least long enough to reproduce). Which organisms fit this type? Most plants and animals! There are many variations on the Type III theme, however, as we shall see.

POPULATION MODELS

To *really* understand the dynamics of a population, we need a method to track and predict how the number of individuals changes through time. For this we will need to use a bit

Type I

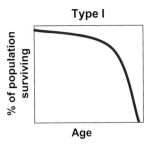

Type II

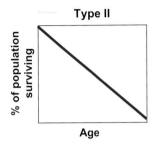

Type III

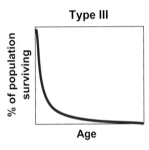

FIGURE 5.12. Three general types of survivorship curves showing the percentage of a population that survives with increasing age. Type I curves are typical for mammals, and Type III curves are typical for insects and weeds. (Figure by M. P. Marchetti.)

of mathematics, but don't despair: we promise to make this as painless as possible.

First we have to figure out just how populations change in numbers. What factors make them grow or shrink? Populations can gain individuals through processes such as birth and immigration and can lose individuals through processes such as death and emigration (leaving a population but not dying). If we put these two factors together into a simple equation, we have the following:

# of individuals gained – # of individuals lost

= change in population number

The change in population number is also referred to as the *population growth rate,* abbreviated by the symbol *r,* so our formula becomes the following:

$$gains - losses = r$$

If you think about it, you may recognize that the population growth rate (*r*) is going to vary among species. For example, *r* will be larger for houseflies than it is for elephants. Based on this simple parameter *r* (population growth rate) we can build a model of population growth called the *exponential growth model.* The model looks like this:

$$G = rN$$

In this formula we already know that *r* stands for population growth rate (i.e., gains minus losses). The *N* stands for the number in the population when we start, and the *G* stands for the change in the population (or growth) over time. The way we work with the equation is simple. First we have to plug in some numbers and calculate *G.* This means we need to know both the population size when we start and the population growth rate.

For our example, let's look at a hypothetical population of cats, where we start with 10 cats (*N*), with a population growth rate of 2. If we plug these numbers into the equation we get *G* = 20, meaning we have a gain of 20 cats. Now comes the fun part. In round two we make the simplifying assumption that our original 10 cats died and we are left with their 20 offspring. So 20 becomes the new value for *N* while the growth rate remains the same, and we recalculate a new *G.* In the third year we get *G* = 40. If we repeat this whole process 10 times, we get the numbers shown in Table 5.1. If we take these numbers and plot them with year (time) on the *x*-axis and population size on the *y*-axis, we will have a graph like the one in Figure 5.13. This kind of population growth, called *exponential growth,* shows a rapid increase in population

TABLE 5.1
Exponential Growth of Cats

Year	1	2	3	4	5	6	7	8	9	10
Population size	10	20	40	80	160	320	640	1,280	2,560	5,120

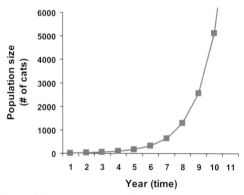

FIGURE 5.13. Exponential population growth in a hypothetical cat population. Note the rapid rise in population size in just a few years and the resulting J-shaped curve. (Figure by M. P. Marchetti.)

size over time. Exponential growth is typified by a J-shaped curve. If we were to keep going with the calculations, we would very quickly find the numbers becoming gigantic, and in fact they eventually approach infinity.

We encourage you to play around with this equation a bit. Try using different values for N but the same value for r, or try using various values for r with the same initial population size. Make sure when you do these calculations you carry them out for at least 10–20 years. What do you find in every case? The answer is that you find both a similar graph (J-shaped curve) and a similar phenomenon where after a while the population size explodes to giant numbers. As we saw in Chapter 2 with cod and rabbits, nature does not seem to allow huge population explosions all the time, so our current model of population growth is clearly inadequate. We need a model that is more realistic and does a better job of showing what happens in nature.

LOGISTIC GROWTH MODEL

We know that plant and animal populations in the natural world don't continue to grow to infinity (but see Figure 5.14); growth generally slows down and eventually levels off or declines. The next model we will look at is a simple addition to the exponential growth model above. We have to add on some term that represents the fact that, in nature, populations have physical limits imposed on their growth. Think about our cat example above. Let's say that you had a South Sea island to yourself and you started raising cats as described. What would eventually happen that would put a halt to the cat population growth? Eventually there would not be enough food for the cats, so they would start to starve and be unable to reproduce. Thus, a shortage of food means no more kittens, which

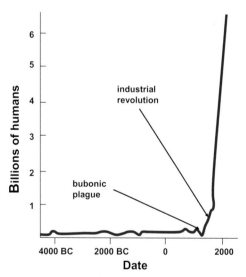

FIGURE 5.14. Estimated population size of human beings (*Homo sapiens*) through history. Note that this graph looks very much like a typical exponential growth curve. (Figure by M. P. Marchetti.)

means no more population growth. We need some term to represent the physical limit or natural capacity of the environment. This value we call k, or the *carrying capacity*, which can be thought of as the amount of resources available to support population growth.

How do we put our new k term into the equation for exponential growth? We have to rig it a bit so that our k term becomes a discounting term, meaning that it can reflect the fact that when a population gets near its carrying capacity, things start to get crowded and population growth will slow down. Our new equation looks like the following:

$$G = rN \frac{(k - N)}{k}$$

Notice that this is just our exponential growth model with a $[(k - N)/k]$ term added. Let's take a look at how this new term works by initially setting your island's carrying capacity to 100 cats (i.e., $k = 100$). If the population size is very low (e.g., $N = 1$), then the $(k - N)/k$ term becomes

$$\frac{100 - 1}{100} = \frac{99}{100} \sim 1$$

If we multiply any number by 1, what do we get? We get the number back. So if we multiply the term rN by 1, we get rN. Thus, when the population size is very low, our new equation acts like our old equation. Now let's look at it when the population is very big. If our population contains 99 individuals then our $(k - N)/k$ term becomes

$$\frac{100 - 99}{100} = \frac{1}{100} \sim 0$$

If we multiply any number by 0 what do we get? We get 0. So if we multiply the term rN by 0, we get 0—or no growth. When the population gets very close to the carrying capacity, the growth of the population ceases, which is pretty interesting.

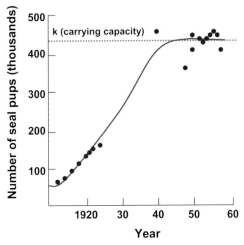

FIGURE 5.15. Northern fur seal (*Callorhinus ursinus*) population growth on St. Paul Island of the Pribilof Islands, Alaska (grey dots). The solid line is the predicted population growth for these seals using a logistic growth model: note its S-shaped curve. The estimated carrying capacity (*k*) for this population is depicted with a dotted line. (Modified after figure from Fowler, C. W. 1990. Density dependence in northern fur seal *(Callorhinus ursinus)*. Marine Mammal Science 6(3): 171–195.)

If we start with any numbers and run our new logistic growth model through a whole bunch of years, we will get a graph like the solid line in Figure 5.15. Notice that the graph looks like the exponential growth model for the first few years, but eventually growth slows down and levels off at the carrying capacity. This line on the graph can be described as S-shaped, or sigmoidal, and it shows an environmental limit to growth, which is common in nature. Thus, most populations exhibit a growth pattern similar to the logistic model.

The data points in Figure 5.15 show a fur seal population on an island in Alaska where they were introduced around the turn of the century. Notice that the actual numbers (the points in the Figure) fit the prediction (the solid line) from the logistic growth model pretty well, but not exactly. But we know the population size, the rate of growth, and the carrying capacity, so why don't the actual numbers of fur seals fit the model? The answer lies in the fact that two of our variables, r and k, are just that: they are variable and do not stay the same over time. Natural environments vary widely over time (think droughts, wet years, and volcanic eruptions). During some years there is a lot of food, so the carrying capacity may go up. Other years are bad for reproduction, so the growth rate goes down. In some years both things happen simultaneously or neither happens. Nature is variable, and the environmental conditions that place limits on population growth are also variable and will certainly change over time. This variability causes populations to sometimes overshoot their carrying capacity. When this happens, there are too many individuals for the environment to sustain and the population will start to decrease. When the population gets below the carrying capacity, it will usually start to increase again, and this adjusting process will continue to happen repeatedly over time.

OVERHARVESTING

We will see versions of these growth models multiple times in coming chapters because these simple mathematical formulas are actually used quite frequently in conservation efforts for wildlife populations. Wildlife biologists or resource managers often want to regulate harvest in some portion of a population: for example, deer for hunters, or salmon for fishers, or wild mushrooms for collectors. Yet the managers often face a dilemma: how can they tell how many deer, salmon, or mushrooms the harvesters should be allowed to take? One way to solve this is to use population growth models very similar to the ones above to predict what will happen if you harvest some fraction of the population. This type of management, where someone is managing for a problem before it becomes a problem, is unfortunately not typical, and many harvested populations have seen disastrous population crashes as a result.

One classic example showing the lack of proper management is the Peruvian anchoveta (anchovy) fishery during the 1960s and 1970s. Figure 5.16 shows anchoveta catches in the millions of metric tons from 1955. During the first few years of the fishery, catches seemed limitless and more boats continued to join in a money-making frenzy. The harvests of anchoveta kept increasing because there was little or no management of the number of fishers or limits to their catches, even when populations were low. By 1972 the fishers had harvested too many anchoveta and the population crashed. Later ecologists, using data from the catch records and models similar to those above, were able to back-calculate the *maximum sustainable yield*, or the number of anchovies that could have been harvested without causing the population to crash. We can see from the graph that the fishery overshot the maximum yield on four of the five years preceding the crash.

Amazingly, with the closing of the fishery for a few years, combined with the return of ocean conditions that favored high growth rates, the anchoveta population returned to high levels, and as a consequence the fishery also resumed. Unfortunately, despite the history of collapse and economic hardship, there are signs the Peruvian fishery is again being set up to collapse, as a result of poor regulation and the high demand for fish meal to feed chickens and farm-raised salmon (the ones generally available in our supermarkets). After the next collapse, the population may not recover, as has happened to other fisheries. Overfishing not only causes economic hardships; it also causes changes to the entire ecosystem of which the anchoveta are part, so populations of pelicans and other seabirds also collapse, as do populations of predatory fishes, such as mackerel. Ironically, Peru is a country in which a significant percentage of the human population have diets deficient in protein, yet most of the anchoveta catch is shipped to other countries.

One of the major conservation lessons from population ecology is that understanding how

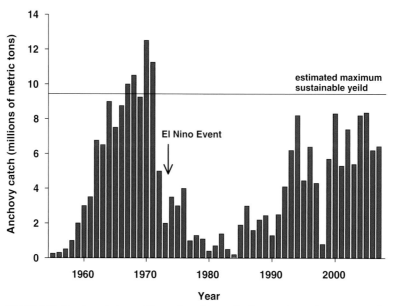

FIGURE 5.16. Yearly catch data (millions of metric tons) for the Peruvian anchoveta (*Engraulis ringens*) population since 1955. Note the extreme drop in catch numbers following an El Niño year in 1973. Conservation scientists estimated the maximum sustainable yield for this population (calculated after the 1973 crash), which is depicted as a solid line on the graph. Note that the anchoveta population recovered to some extent in the late 1980s and 1990s but unfortunately continues to wildly fluctuate, indicating an inherently unstable population. (Redrawn with permission from Glantz, M. H. 1981. Consideration of the societal value of an El Niño forecast and the 1972–1973 El Niño. pg. 449–476. In Resource management and environmental uncertainty: lessons from coastal upwelling fisheries, volume 11 in the Wiley series in Advances in Environmental Science and Technology. Eds. M. H. Glantz and J. D. Thompson. Wiley Interscience, New York.)

populations work, especially how they respond to environmental change and human manipulation, is crucial for designing conservation strategies. All populations have their limits, and these limits are determined by the capacity of an ever-changing environment. Thus, conservation must involve understanding not only population dynamics of different species but also how these dynamics change as the environment changes.

CONCLUSIONS

Animals and plants live in a variable world to which they must continually be able to adjust both as individuals and as populations. Thanks to the processes of evolution and natural selection, organisms have developed a remarkable ability to thrive in this variable world. Yet the world is not overrun (usually) with rabbits,

bighorn sheep, and desert iguanas. As we have seen, both individuals and populations have limits, and exceeding those limits can result in deaths, population crashes, and further environmental change. We humans, however, tend to think of ourselves as immune to such population or ecological limits. In reality we are not.

FURTHER READING

Clover, C. 2006. The end of the line: how over fishing is changing the world and what we eat. University of California Press. Berkeley. *If the anchoveta story intrigues you, read this book for a bigger picture.*

Krebs, C. 2008. The ecological world view. University of California Press. Berkeley. *This is a particularly accessible example of the many ecology texts in existence that can provide additional background on the subject matter considered here very briefly.*

Kurlansky, M. 1997. Cod: biography of a fish that changed the world. Penguin Books. New York. *The cod, once one of the most abundant fish in the Atlantic Ocean, played a major role in the development of the modern world. This is the story, in readable form, of the cod through history, and it helps to explain how the cod populations collapsed.*

Pavlik, B. M. 2008. The California deserts: an ecological rediscovery. University of California Press. Berkeley. *A nice example of the many well-illustrated books published in recent years on particular habitats or regions. Read this, and you will acquire a good understanding of the habitats of kangaroo rats, desert iguanas, and desert bighorn sheep (and many other organisms).*

6

Ecology

COMMUNITIES AND ECOSYSTEMS

In the last chapter, we described two of the fundamental units of ecology, but we still only talked about single species. Clearly, species do not live in isolation; there are many other organisms that are part of each and every species' environment. When we move from the ecology of single species to the ecology of how different species relate to each other, we enter the world of biotic communities. Figure 6.1 illustrates how one might represent these different levels of ecological organization. But communities are not the final level of organization. Communities interact with their larger environment and with each other: this is the realm of ecosystem and landscape ecology. In this chapter, we provide an introduction to these higher levels of ecological organization and show how they can help us to form a coherent picture of conservation science.

COMMUNITIES

A biotic *community* can be defined fairly simply as *a group of species that interact with each other.* At its most basic, this means that anytime you have more than one species to deal with, you are looking at some part of a community. All of the following are examples of natural communities: (1) the trees, shrubs, flowers, lichens, mosses, fungi, insects, amphibians, mammals, and birds that live together in a pine forest; (2) the algae, invertebrates, fishes, birds, and mammals that live in or on a freshwater lake; (3) the corals, worms, mollusks, sea anemones, starfish, sea urchins, and fishes that live in or around a tropical coral reef. The point is that species do not exist on their own. They are constantly in contact with other species and typically have to make some adjustments because other species are present. It is these interactions, the ones among different species, on which community ecologists focus their research.

So if a community is a group of interacting species, what do we mean by "interact," and how do these interactions work? If you think about it, you can see that species in a community have either positive interactions with most other species (e.g., pollination) or they have negative interactions (e.g., being eaten). Ecologists generally recognize four broad classes or

| Individual | Population | Community | Ecosystem |

FIGURE 6.1. Four levels of biological and ecological organization: individual, population, community, and ecosystem. Individual ecology deals with how organisms adapt to their environment. Population ecology is the study of how a group or population of an individual species grows or declines. Community ecology focuses on how populations of different species interact together. Ecosystem ecology is concerned with how biological (biotic) and nonliving (abiotic) forces interact to produce local and global ecosystems. (Figure by M. P. Marchetti.)

types of interactions among species: competition, predation, mutualism, and parasitism. We will examine each of these in turn, followed by a discussion of the process of succession, or how communities change through time.

COMPETITION

We begin by defining competition as an interaction where individuals of one species suffer reduced fitness or reproduction as the result of an interaction with individuals of a second species. In other words, two species meet, and at least one of them is harmed by the encounter. In order for competition to occur, scientists have recognized that three conditions need to be met. First, the two species must use or depend on some common resource. Second, this resource must be in short or limited supply. Finally, as a result of the encounter, individuals of one of the species must suffer reduced survival or reproduction.

This is all well and good, but how does this translate into the real world? For starters, what do we mean by a common resource? Or, to ask a slightly different yet related question, what types of things could two species compete for? This question is not too difficult to answer if we use our own lives as a starting point. What might you compete for with another species? One obvious answer is food. Two species might eat the same type of food. For example, you might compete

with rabbits or deer or crows or insects for the vegetables you grow in your garden. This is pretty easy to see in terms of animals, but what about for plants? Most plants don't eat food in the sense we generally use the word, but they do need nutrients from the soil (think plant fertilizer) as well as water and sunlight in order to make food by photosynthesis, so they could compete for these types of resources. Another resource you might compete for is space; space to live in, space to grow in, space to find food, or space to hunt. For animals, this could be a den or a cave or a tree hole, a portion of a forest, a section of stream, or some other defined territory. For plants, because they don't generally move, this could literally be space to germinate and/or space to grow and spread out. Animals also sometimes compete for reproductive sites or territories such as secluded spots in a tree in which to put a nest or sections of a beach where they can lay their eggs. The types of resources over which competition can occur are almost limitless.

Yet there must be a limited supply of the disputed resource. Why is this necessary? A simple thought experiment should answer this. Let's say that we lock you in a huge room for three days with a vicious dog and the only food available is pepperoni pizza, which both of you like to eat. If there is only one pizza in the room, you are clearly going to compete with the dog over the pizza (we might assume you would win with your superior intellect, but the dog could be really aggressive and hard to beat). But if there were hundreds of pizzas in the room, then you and the dog could happily go about snarfing your pizza for days on end without the need to interact over the food supply (read the novel the *Life of Pi* by Yan Martel for an interesting version of a similar scenario). The bottom line here is that a large supply of a resource equals little or no competition, whereas a limited supply of that resource creates competition between species.

Despite the fact that competition can take many forms, the potential range of outcomes for competition is rather limited. If you think about it, either one species can win the competition or the two species can somehow divide up the resource. First, let's look more closely at what we mean by one species winning. Ecologists have a name for this phenomenon: *competitive exclusion*. In this situation, ecological and evolutionary theory both suggest that a species less able to compete will either change its habits (adapt) or die out; as a result, one of the two species excludes the other.

We can see an example of this phenomenon using two species of paramecium. You may remember from high school biology that paramecia are microscopic animals that live in pond water. What you might not have known is that there are many different species of paramecium that can live in the same pond. In a classic experiment from the 1930s by Georgii F. Gause, two species of paramecia, *Paramecium caudatum* and *Paramecium aurelia*, which both eat the same food (single-celled algae), were grown both in separate jars and then together in the same jar. Figures 6.2A and 6.2B show how the populations of the two species grew when they were in jars by themselves. You can see that, for both species, the populations increased rapidly up to a certain point and then leveled off. (You should remember from the previous chapter that this kind of population growth is called logistic growth and that the population ceiling is called the carrying capacity, k). Interestingly, something very different occurred when the two species were grown together in the same

jar (Figure 6.2C): *P. aurelia* was better at collecting algae, causing the population of *P. caudatum* to die out, and therefore competitively excluding the species. Competitive exclusion can happen with larger organisms (plants and animals) as well and is thought to be one of the major forces in the process of natural selection we described earlier.

Of course, if exclusion were the only outcome of competition, it is likely the planet would be populated by only a few species, so we know that there must be another possible outcome when two species compete over a resource. This second outcome is often called *resource partitioning,* which basically means that the two competing species somehow divide up the resource. A great example of this comes from the work of one of the great scientists in modern ecology, Dr. Robert MacArthur. Dr. MacArthur was intrigued by the fact that in one pine tree he would sometimes find up to five species of warbler (a small bird) feeding. All of these warbler species essentially eat the same thing, namely insects, so it seems as if they should compete. Why did Dr MacArthur find so many species in the same tree when the process of competitive exclusion would predict only one? After many days of studying the birds closely, he realized that the warblers didn't see the tree as one whole thing. They instead divided up the resource (i.e., space, which in this case is the tree) and foraged for insects in different areas of the canopy. For example, the yellow-rumped warbler foraged in the bottommost branches of the tree, the bay-breasted warbler fed in the middle branches, and the Cape May warbler gleaned insects only from the top-most area. In this way, the birds actively avoided competition and could coexist while still using the same apparent resource.

A second good example of this also involves birds, but one of the species we have already been introduced to, the medium ground finch *(Geospiza fortis)* from the Galapagos Islands. In Chapter 2, we saw that beak size was an important element in the feeding of these birds, but what we didn't mention was that

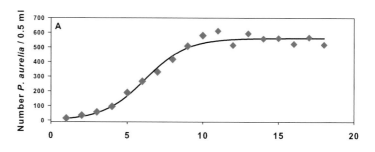

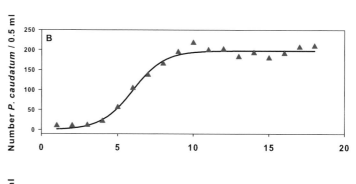

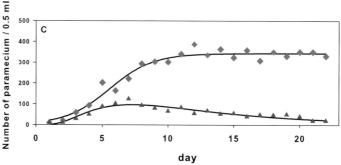

FIGURE 6.2. The famous experiments by Georgii Gause showed that, when populations of two species of paramecia (*P. Aurelia*, graph A, and *P. caudatum*, graph B) are grown alone, they exhibit typical logistic (S-shaped) growth, and both reach a carrying capacity (*k*). But when they are put together in the same environment and forced to interact (graph C), *P. aurelia* is the better competitor for the common food resource and, as a result, drives *P. caudatum* extinct. (Data for the figures were taken from Leslie, P. H. 1957. An analysis of the data for some experiments carried out by Gause with populations of the protozoa, *Paramecium aurelia* and *Paramecium caudatum*. Biometrika 44(3/4):314–327.)

there are a number of different finch species found around the Galapagos, all of which compete for the same resource, namely seeds. The Galapagos is a group of islands, some of which are very tiny and house only one species of finch and some of which are considerably larger and contain two or more species. When we examine the average beak size of *Geospiza fuliginosa*, the only species that inhabits the Los Hermanos islets, we find considerable overlap with the average beak size of *G. fortis*, which solitarily inhabits Daphne Major Island (Figure 6.3). This suggests that the two species should strongly compete when we find them together. Interestingly, on Santa Cruz Island, where these species co-occur, we find the average beak size for both species has shifted; *G. fortis* have slightly larger beaks whereas *G. fuliginosa* have shifted toward a smaller beak size. This example shows how competition can act as a force driving evolutionary change similar to the way we saw that variation in the environment (e.g., the weather on the Galapagos) can cause evolutionary change.

Regardless, competition is an important ecological factor that shows up time and time again in the study of conservation, particularly when we turn to the study of invasive species in a later chapter. It's worth noting here that theoretical ecologists and conservation practitioners have developed mathematical models of competition that involve simple modifications of the

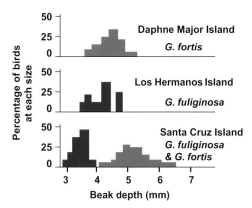

FIGURE 6.3. In the foundational work by Peter Grant on the Galapagos finches, Dr. Grant and colleagues found that, on islands that contained only one finch species (Los Hermanos and Daphne Major), the average beak sizes of the two species converged. Yet when the same two species were found together on certain islands (Santa Cruz Island), the beak sizes were measurably different. This is a good example of resource partitioning between competing species. (Modified after figure from Grant, P. R. 1986. Ecology and evolution of Darwin's finches. Princeton University Press. Princeton, NJ.)

logistic growth equation we examined in the last chapter.

Logistic Growth Equation

$$G = rN \frac{(k - N)}{k}$$

Two-Species Competition Equations

$$\text{Species 1: } G_1 = rN \frac{(K_1 - N_1 - \alpha_{12} N_2)}{K_1}$$

$$\text{Species 2: } G_2 = rN \frac{(K_2 - N_2 - \alpha_{21} N_1)}{K_2}$$

These models can often help predict the outcome of two-species interactions and assist in the management of invasive species. The point here is not to dwell on the details of this competition model but to highlight the fact that fairly sophisticated conservation work can be derived from simple tools.

PREDATION

A second class of community interactions involves one species using another species for food and is called *predation*. By this definition,

virtually every living thing is either predator or prey, and often both. When a bear eats a salmon, we call the bear the predator and the salmon the prey. How many examples of this type of interaction can you think of? It's worth trying to be a little creative here. Predation can actually take many forms and is given various names depending on the participants. The typical mammal eating another animal scenario is pretty clear, but when a deer eats leaves off a shrub, is this predation? The answer is yes it is, but it is a special case because the shrub usually does not die in the process. Many grazers such as cows, horses, sheep, deer, numerous insects, and many others that consume plant material are also called *herbivores,* and the activity of eating plant tissue is called *herbivory.* We can see that herbivory is generally a type of grazing, although if the entire plant were to be consumed, then it might justifiably be called predation.

Plants aren't the only organisms grazed; animals can be grazed upon as well. A starfish may loose an arm to a predator but can grow a new one and so does not always die in the interaction. In a similar way, you and I could be said to be "grazed upon" by female mosquitoes when we get bitten, because they are taking some of our tissue (blood) to produce eggs. This latter type of predation can be called *micropredation* because the prey is much larger than the predator and usually survives the experience. Other examples include lampreys, which attach to large fish and suck their bodily fluids, and vampire bats, which lap the blood of large mammals after creating a neat incision. Micropredation is sometimes regarded as *parasitism* (but see next section). Clearly, the realm of predation or using other organisms for food is a big, encompassing topic.

The process of predation is more clearly defined than competition, in that if some critter is being eaten, it is pretty clearly predation. But ecologically, what are the possible outcomes between a predator population and its prey population? One possibility is that the predator exterminates its prey. We can see an example of this in another famous experiment by G. F.

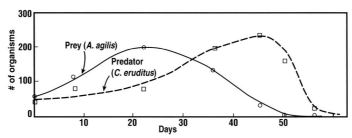

FIGURE 6.4. In a classic experiment, Georgii Gause and his colleagues explored the interaction between a predator (a mite called *Cheyletus eruditus*) and its prey (a different mite, *Aleuroglyphus agilis*). This figure shows population growth for the two species when they were grown together. Note that 30 days after the predator was introduced, the prey population began a rapid decline and was eventually driven extinct, which in turn led the predator population to be extirpated as well. (Reprinted with permission from Gause, G. F., N. P. Smaragdova, and A. A. Witt. 1936. Further studies of interaction between predators and prey. The Journal of Animal Ecology 5(1):1–18.)

Gause, this time using two species of mites (tiny arthropods related to spiders). In this experiment, he examined a mite named *Aleuroglyphus agilis* and its predator, another mite called *Cheyletus eruditus*. When he grew a population of *A. agilis* before adding a predator, it showed a typical logistic (or S-shaped) growth curve. But when he added a population of its predator *(C. eruditus)*, the prey population decreased and quickly declined to zero (Figure 6.4). From looking at this figure, you might also note what happened after *A. agilis* disappeared: the predator population also quickly declined and eventually died out. If predators always exterminated their prey, then we would have very few organisms on the planet, so again something else must be happening in the majority of predatory interactions.

One answer is *population cycles,* where the populations of both predator and prey cycle from low to high and back again. If you think about it, this makes some sense. If the predator population is eating its prey very quickly, then eventually there will not be enough prey to go around, and some predators will die of starvation. As more and more predators die, the pressure on the prey population is reduced, and prey numbers start to increase because there are not many predators around. Over time, this creates cycles of boom and bust for both predator and

prey. A classic example of this is seen with the Canadian lynx *(Lynx canadensis,* a kind of wild cat) and snowshoe hare *(Lepus americanus)* populations at the turn of the last century. Both of these animals were hunted and trapped for their fur, and the early trading companies kept good records of the number of pelts the hunters brought in each year, with more pelts indicating a larger population size. When these data are graphed over time, we see a very clear pattern (Figure 6.5). The lynx (predator) population always peaks a year or two behind the high point in the hare (prey) population, which is what you might expect.

A curious observer of the lynx-hare population cycles might ask, Is the lynx driving the hare population, or is the hare driving the lynx population? An enormous research effort has attempted to answer this very question. A multi-decade, multimillion dollar study in the far Canadian north was spearheaded by the ecologist Charles Krebs. Krebs and colleagues noticed that there was more than one type of predation involved in this interaction; the hare, too, are predators, but on grasses and shrubs. They wondered if the hare population was responding to increases and decreases in its food supply (caused by more or fewer hares, as well as by climate fluctuations), which in turn would

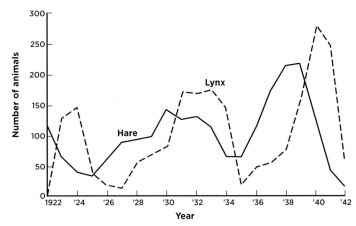

FIGURE 6.5. Predation pressure often leads to population cycles in both predator and prey, as can be seen by looking at the cycles of Canadian lynx *(Lynx canadensis)* and its principal prey, showshoe hare *(Lepus americanus)*. As the numbers of lynx increase, there is a decrease in hare numbers, but as hare become scarce, the predator population will also eventually decrease. When the lynx population is very low, there are not many predators around to keep the hare population in check, and the hare population grows rapidly. (Reprinted with permission from Bulmer, M. G. 1974. A statistical analysis of the 10-year cycle in Canada. The Journal of Animal Ecology 43(3):701–718.)

drive the lynx population. Or, was the lynx driving all the other cycles? After many years and an incredible amount of effort, the results suggest that, in fact, both are occurring. The hare population responds to changes in its food supply, while at the same time the hare population also responds to changes in its predator population. Both types of predation seem to be driving the cyclic nature of these animals.

Another answer to the puzzling coexistence of predators and prey is found in the idea of *optimal foraging*. This simply means that predators try to maximize the amount of prey they capture with the least amount of effort. For example, if a bird-eating predator, such as a Cooper's hawk, has a hard time finding a preferred bird prey (e.g., magpies), then it will switch to other kinds (e.g., flickers). Likewise, you may prefer to eat steak but are likely to eat carrot sticks if that is all that is available. In fact, most predators feed on many kinds of prey, so they are relatively immune to cycles of one kind of prey and vice versa.

It is worth noting that in order to predict predator and prey cycles, theoretical ecologists have also developed mathematical equations derived again from our simple logistic growth model in the previous chapter.

Logistic Growth Equation

$$G = rN \frac{(k - N)}{k}$$

Predator Prey Equations

$$\text{Prey: } G = rN \frac{(1 - N) - aNP}{K}$$

$$\text{Predator: } G = caNP - dP$$

Please note the similarities among the models. Again the details of these models are not important for us; instead, we want to recognize that these equations are extremely useful for the management of animal populations, especially those that are harvested (deer, salmon, pheasant, geese, etc.), as well as for assisting conservation and restoration efforts.

PARASITISM

Every organism is host to multiple kinds of parasites, even you, so parasitism is the commonest kind of interaction among different species.

"True" parasites are multicellular organisms that live on or in a host organism, depend on the host for their survival (especially for nutrition), and ultimately harm the host. Singled-celled microorganisms and viruses that give us diseases are also parasites, although we usually think of them separately as *pathogens*. Pathogens, of course, are among the most studied of organisms, largely because humans want to find ways to control them and to keep them from killing large numbers of us in epidemics. Pathogens are just as important to non-humans and are a major cause of death. Consequently, there is always an evolutionary race between pathogens and the animals they infect. Ideally, the pathogen should not kill its host, at least not right away. It is believed that there may be some benefit to the pathogen if the host develops some resistance to the pathogen's effects. The bubonic plague was originally a disease of rodents, for which they evolved resistance, but when it jumped to humans, it caused huge mortality because humans lacked resistance. The plague pathogen still persists in rodent populations, waiting to jump to us again.

Parasites also find it advantageous to keep their hosts alive because their own lives may depend on it. Thus, a tapeworm lives in the intestine of its host, obtaining nutrition in various ways and shedding eggs into the intestine, which are carried outside in the feces of the host, where they spread to other hosts. Some parasites, however, sacrifice their hosts to complete their life cycles. For example, there is a trematode parasite of sticklebacks (a fish) that lives as a juvenile in the fish but has to infect a bird to become an adult. When the trematode is ready to infect a bird, it moves into the brain of the stickleback and causes it to swim erratically near the surface of the water where a fish-eating bird is likely to eat it. Once eaten by the bird, the parasite can then complete its life cycle inside the new bird host.

The most common effect of both parasites and pathogens is to reduce the ability of their hosts to survive or reproduce. This is because a host that diverts large amounts of energy to sustain a parasite has less energy to sustain itself.

This makes it more likely to lose in a competitive interaction or more likely to be eaten by a predator. Bird chicks in a nest heavily infested with ticks will grow more slowly and are less likely to survive than chicks with few ticks. Thus, avoiding tick infestations is one reason why birds rarely reuse an old nest.

MUTUALISM

Our fourth major class of species interactions has been the least studied of the four, but this fact should not lead us to the belief that mutualisms are somehow less vital to the planet. In fact, an appreciation of the ubiquity of mutualisms and their importance for the maintenance of biodiversity and human well being is growing rapidly among ecologists, conservationists, and resource managers. A *mutualism* can be defined as an interaction between individuals of two species where both parties benefit. In a sense, competition, predation, and parasitism are negative interactions for at least one of the parties involved, whereas mutualisms are good for both participants. Can you think of any interactions in nature where there is a net gain for both individuals? There are some easy ones that should spring to mind, particularly the close relationship between bees and flowers. Here, the two parties need different things; the bees need nectar or food, and the flowers need to be pollinated in order to reproduce. While bees are collecting nectar inside a flower, the plants make sure that the bees will be covered with pollen (male gametes). The pollen is then transferred to the next flower on the bee's nectar-collecting route, allowing for cross-fertilization (and thereby increasing the amount of variation in the flower population). Both parties in this exchange benefit; the bees get fed, and the flower gets to reproduce.

There are many other fascinating examples of mutualism in nature, including one involving common coral reef fish called cleaner wrasses. In this case, the mutualism exists between one species (the wrasse) and many other fish species. The wrasses set up a "cleaning station" in some prominent place on the coral reef and exhibit a swimming display or show some bright colors to attract larger fish. A large fish, which could easily eat a small wrasse, will sit quietly while the wrasse moves over it, feeding on parasites and damaged tissue. Sometimes this will actually involve the wrasse swimming into the jaws of the larger fish and cleaning parasites inside its mouth, without the larger fish eating the wrasse. The wrasse gets a stable supply of food, and the larger fish gets cleaned and freed from diseases; therefore, the whole interaction is beneficial to both parties. Naturally, there are studies that show there is also a dark side this beautiful picture: cleaner fish will sneakily eat the eggs of fish defending nests while cleaning the defender. Fish will also come to be cleaned even if they don't have any parasites; apparently, they just like the way cleaning feels!

Currently, one of the fast growing areas of mutualism research involves a positive relationship between fungi and plants. There are many species of fungi living in the soil, some of which produce fruiting bodies (i.e., mushrooms) and some of which do not. Some soil fungi called mycorrhizae will enter into a mutualistic relationship with the roots of plants. The plant benefits because the mycorrhizae are very good at acquiring essential nutrients from the soil, such as phosphorous, and will "trade" with the plant. The plant, for its part, is much better at getting carbohydrates (i.e., sugars or food) through its ability to photosynthesize and so will trade this with the fungi. The plant gets essential elements, the fungi gets food, and both benefit and grow substantially better. This particular mutualism is extremely important for conservation work involving the restoration of native plants. When trying to restore native vegetation to a degraded site, the restoration ecologist wants to give the native plants every possible advantage so they can outcompete the exotic vegetation (see Chapter 12 for more information about restoration). If the correct mycorrhizae are included in the soil when the natives are planted, then the entire restoration effort is generally more successful. Understanding mutualisms can therefore help us to more effectively repair some of the damage to our forests and grasslands.

SUCCESSION

Besides the four basic types of community interaction, there is another aspect to community ecology that deals with the element of time, particularly how communities change over time. As a community gets older, there is a more or less predictable series of changes that occur, and this process of community change is called *succession*. In the successional process, the community generally goes from simple to complex in terms of both the number of species and the connections and interactions among them. This is easy to see if you have ever watched someone's yard suffer from neglect over a number of years. A typical yard in an American suburb will generally contain one species of grass, some trees, and some shrubs. If a yard is not maintained, then the grass will be invaded by other grasses, forbs (small non-woody plants), and shrubs. If the yard is neglected long enough, it may eventually come to resemble whatever community is native to that region (e.g., forest, desert scrub, meadow). The same thing happens when an agricultural field in the eastern United States is left fallow (unmaintained) for a number of years. Eventually, the agricultural field is replaced by a forest. A similar process happens across all communities; they change slowly and somewhat predictably through time.

Succession comes in two distinct but related flavors, one called primary succession and the other called secondary succession. Primary succession occurs when the process starts from an absolute blank slate: a location where there is no life at all. Can you think of any places on the planet that might fit this description? If you have ever visited the big island of Hawaii and gone to Volcano National Park, then you may have stood on one such place. Recent lava flows are absolutely bare substrate; you will not find a blade of grass or a shrub or even a seed anywhere on the new lava rock. All the Hawaiian Islands were formed this way, and it's amazing

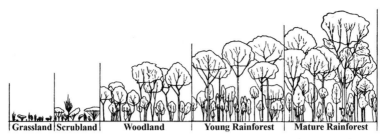

| Grassland | Scrubland | Woodland | Young Rainforest | Mature Rainforest |

FIGURE 6.6. Schematic drawing of the successional process, showing changes to a clear-cut rainforest plot through time. Succession theory suggests that over an extended period, a plot of land will proceed through a predictable series of changes in both the plant and animal communities. (Redrawn with permission from Beard, J. S. 1955. The classification of tropical American vegetation-types. Ecology 36(1):89–100.)

to see how species rich the islands eventually became. Another location for primary succession is found after a glacier retreats and leaves bare rock and gravel. The first colonizers of this barren substrate are small plants such as lichens and mosses that actually break down the rock and begin to make soil. Once these plants create soil or otherwise stabilize the area, they are replaced by grasses and shrubs, which shed leaves and add to the depth of soil. The shrubs come to be replaced by trees and eventually an entire mature dense forest (Figure 6.6). This process, going from bare rock to complex forest, is called primary succession, and it usually takes hundreds to thousands of years, depending on the location and scale of the succession.

Secondary succession is a similar process, but one that does not start from the same bare starting point. We call succession secondary when the process begins with some type of disturbance, but following the disturbance, there is still living material (i.e., roots, seeds, fungi, etc.) remaining in the area. For example, when a wildfire burns through a forest, if it is hot enough, it may kill off all the living plants aboveground. But if you visit such a place the year after a fire, you will find grasses and shrubs and even small trees beginning to grow. The fire's heat may have killed all the plants, but it doesn't kill all the seeds and roots and tubers that are living just a few inches below the surface. Many plants have a considerable amount of tissue below ground, and some can quickly re-sprout even if

you remove all the aboveground tissue. Secondary succession typically follows the same path as primary succession, but it essentially has a head start in time, so it doesn't take as long. Other types of natural disturbance besides fire can initiate secondary succession; floods, droughts, hurricanes, and landslides all can begin the process. Secondary succession is also not limited to natural disturbances. Human-caused alterations to a landscape, such as from a bulldozer, clear-cut logging, or clearing for agriculture, can also start the process.

When we say that succession is predictable, we do not mean that scientists can tell exactly what year the grasses will invade the mosses and lichens, but instead that the general pattern of moving from simplicity to more complexity is consistent over time and space. We can predict the general membership (which species will be present) in the final community by knowing what type of community is found in the area or is adapted to that environment. For example, if a wildfire burns off a pine forest in the northern Rocky Mountains, it will never grow back as a tropical mangrove forest; it is just not possible. After years of successional changes, the burned Rocky Mountain area will come to closely resemble the original pine forest that was found there, although it is likely to have a somewhat different composition of tree and shrub species, due to random processes or changes caused by erosion, climate shifts, and other factors.

TABLE 6.1

Life-History Traits of Some r-Selected and k-Selected Species

TRAIT	r-SELECTED POPULATIONS	k-SELECTED POPULATIONS
Age at first reproduction	early	late
Lifespan	short	long
Juvenile mortality rate	high	low
Number of offspring per reproductive episode	many	few
Size of offspring	small	large
Parental care	none	some

Successional processes are predictable in another way, but to explain it, we have to return to some material from the previous chapter on population models. In those models, we identified two important population variables: r, the intrinsic growth rate, and k, the carrying capacity. Ecologists have theorized that the environmental conditions present in a habitat tend to drive the evolution of a species' traits in one of two general directions, called r-selection or k-selection. Typically, r-selected species exploit unused resources and produce many offspring, each with a low probability of surviving to adulthood. In contrast, k-selected species are strong competitors and invest heavily in fewer offspring, each of which has a relatively high probability of surviving to adulthood. There is a whole suite of what are called life-history traits associated with either r-selected or k-selected species, some of which are listed in Table 6.1.

After reading this list, can you think of species that would likely be called r-selected? How about k-selected? Generally, r-selected species are small, fast-reproducing organisms such as mice, rabbits, most insects, grasses, weeds, and many others. K-selected species are generally large and slow-reproducing organisms such as whales, bears, tortoises, redwood trees, pine trees, and many others.

At this point, you may be asking yourself how all this relates to succession? The answer is that the successional process is predictable in terms of the relative abundance of r- and

k-selected species present. In general, early succession is dominated by r-selected organisms, and over time, the community shifts to being dominated by k-selected organisms. So we go from grasses, mice, and insects to shrubs, trees, bears, and spotted owls. One thing to keep in mind here is that the designation r or k is not cut in stone; instead, the terms are relative to the groups you are comparing. For example, among trees, there are very long-lived k-selected trees such as the giant sequoia *(Sequoiadendron gigantium)* that can live more than 3,000 years. There are also more r-selected trees such as sandbar willows *(Salix sp.)* that typically live less than 50 years. Yet when compared with the fast reproducing grasses, both tree species would be considered k selected. In any case, the process of succession generally moves from r to k, and this knowledge can be very important and help inform and prioritize conservation and restoration efforts. For example, it may be fairly easy to restore an early successional grassland community relatively quickly, but eventually, k-selected species may move in, ruining the restoration effort. If, instead, you were to restore areas of successional grassland interspersed with both early and late successional blocks, you could create a larger mosaic of habitats that would likely be self-sustaining and contain all the important species, with the blocks shifting their communities to earlier and later stages through time. In addition, by including multiple successional stages in a restoration project, the entire process of succession could also be restored and protected, which, as we will see, is often one of the goals of restoration work.

ECOSYSTEM ECOLOGY

As we make the move from communities up to ecosystems, we need to begin with a definition of an ecosystem. A good starting definition for an ecosystem is all the living organisms that interact together in a place, combined with all the nonliving characteristics of that place. This definition in a sense brings us around almost full circle, because we began this introduction to ecology by talking about the environment, climate, and weather. We then examined how individuals adapt to environmental conditions, how a population of individuals can behave as a whole, and how groups of populations interact with each other, and now we will explore how communities interface with the world at large. Perhaps it would be wise at this point to actually define what we mean by *ecology*. *Ecology is the study of how organisms interact with each other and their environment.* You can see from this that our definitions are very clearly connected. All of these step-like levels are interrelated, and as a result, an organism's ecology depends somewhat on the integration among all of these factors and forces. In our discussion of ecosystems, we will be highlighting this very real sense of interconnection.

Ecosystems can be studied by focusing our attention on various patterns in nature at this larger scale. We can study patterns of feeding among organisms (i.e., who eats whom), patterns in how energy flows through systems, and patterns in how nonliving material (i.e., nutrients and inorganic matter) moves through an ecosystem. We will look at each of these in turn.

TROPHIC ECOLOGY: WHO EATS WHOM

Populations within a community are bound together by a network of relationships, one of the most important being *trophic, or feeding, interactions.* The word *trophic* comes from the Greek,

trophos, meaning feeder, and so whenever you see this root in a word, it will always deal with some aspect of food or feeding. For example, what do you think the word *autotrophic* means? *Auto* means something like "self," and *trophic* means food, so autotrophs are organisms that make their own food, which are otherwise known as plants (note that there are non-plant autotrophic organisms such as sulfur-reducing bacteria, but we don't need to dwell on this fact here). Autotrophs are also sometimes called primary producers because they are the originating or primary source of all food on the planet. Autotrophs take sunlight, CO_2, and water, and through photosynthesis, they make all the food that the rest of life depends on. If we were to kill off or damage the autotrophic systems on the planet, life as we know it would quickly cease to exist.

Another useful term for us is *heterotrophic.* Given what we have seen so far, what do you think this word means? If we take the common word heterosexual and remember that this describes a person attracted to the other sex, we can see that heterotrophic organisms get their food from other organisms. We humans are heterotrophs, as are some bacteria and viruses as well as most animals. Heterotrophic organisms are also called consumers because they consume each other and primary producers. There is a subset of heterotrophs that get their own name because they get their food from dead organisms. We call these kinds of critters *decomposers,* or *detritivores,* because they take dead organic matter and break it down further. These can include everything from mammals and insects to fungi and bacteria.

One simple way of looking at trophic patterns is to construct a *food chain,* which is *a linear representation of who eats whom.* For example, here is a simple food chain.

grass → grasshopper → meadowlark → hawk

Note that each level in a food chain is given a different name. The first trophic level is the primary producer (grass) or autotroph. The second trophic level is a primary consumer, in this case

FIGURE 6.7. Hypothetical forest food web. Note the number of arrows and the ecological connections between organisms. This web is greatly simplified because it shows little omnivory and only a small role for the processing of organic detritus. (Figure by M. P. Marchetti.)

an herbivore (grasshopper). The third trophic level (meadowlark) is called a secondary consumer and is either an omnivore or carnivore, depending on whether a majority of its food is from animals. The fourth trophic level (hawk) is variously called a top carnivore or a tertiary consumer. Note that this food chain is pretty simple and that the arrows go in one direction, pointing toward the consumers (if we put the food chain in a vertical direction the arrows would go up). Why is this? The arrows represent how energy or food is flowing in a food chain and therefore point in the direction of the energy's path.

You may already notice that a food chain is really an oversimplified cartoon, but perhaps it's not clear how it is simplified. One thing a food chain fails to do is to recognize the fact that meadowlarks do not only eat grasshoppers. They also eat a wide variety of other insects. Meadowlarks and many other organisms eat many different things (called omnivary), and a food chain fails to show this. A second difficulty is in the fact that most (not all) food chains do not indicate that all four of these critters die and eventually are eaten by detritivores, or decomposers. Food chains rarely show this alternative energy pathway, called the detrital food chain. Finally food chains also fail to indicate that, many times, an organism's diet will change drastically during its life. Think for a moment about the food you ate

as a six-month-old baby compared to what you eat now. It was a pretty dramatic change. This type of developmental diet shift is also not generally represented in food chains.

Of course, there is an alternative to a food chain as a model of feeding interactions, which is called a food web (Figure 6.7). A food web provides a better approximation of trophic reality. We can see that this hypothetical deciduous forest food web corrects many of the issues we had with food chains. There are a number of organisms that have multiple arrows, indicating their omnivorous character, and we see that the detrital pathway is also included. Unfortunately, even this food web is still a very long way from representing the true complexity of nature. In fact, it is extremely difficult to construct an accurate depiction of all the trophic interactions within an ecosystem. The late ecologist Gary Polis spent a large portion of his research career assembling the data for one complete food web in the desert southwest, and even he did not get the entire web identified. To give you an idea of how mind-bogglingly complex real food webs are, we want to show you a partial food web for the marine ecosystem off the Atlantic coast of the United States (Figure 6.8). One thing you should notice from this figure, besides observing that our old friend the Atlantic cod is at the center of this web, is the commonness of

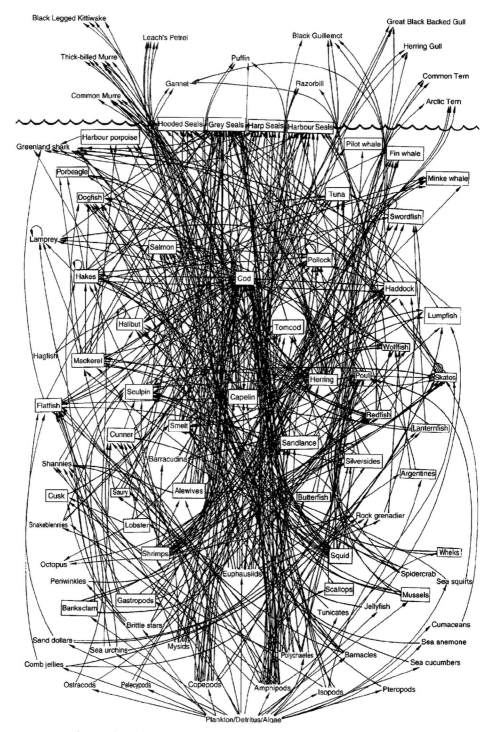

FIGURE 6.8. Atlantic cod *(Gaddus morhua)* food web diagram from the nearshore marine community off the northwestern coast of North America. Note that even this complicated diagram is not a complete food web, as many of the species listed are actually groups containing large numbers of species (i.e., "shrimps," "mussels," or "isopods"). (Reprinted with permission by David Lavigne.)

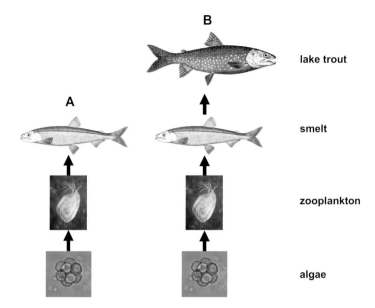

B

lake trout

A

smelt

zooplankton

algae

FIGURE 6.9. Cayuga Lake (New York) food chain: (A) before the addition of lake trout, and (B) after the addition of lake trout. The addition of a top predator (trout) decreased smelt numbers, which in turn increased zooplankton numbers; the zooplankton grazed down the algae that had turned the lake green. This trophic cascade in the food web changed a green lake to one that was relatively clear. (Figure by M. P. Marchetti.)

omnivory. Every player in this web has multiple links to other organisms, meaning that all the critters living in an ecosystem are intimately connected by feeding relationships with a whole host of other species. The so called "web of life" is real, exists everywhere we look, and is incredibly intricate.

You might imagine from looking at Figure 6.8 that removing players (i.e., through extinction) from the food web could have a dramatic impact on the entire structure. Another way of framing this is by asking the question, What happens when a food web is altered? Can we see the impact of altering a food web (by either gaining or loosing a member) in the populations of other species? The answer is that sometimes you can observe an effect. One area where ecologists see such an effect is called a *trophic cascade*. The term *trophic cascade* was coined in the 1970s to suggest that changes in a food web propagate, or cascade down, to lower trophic levels.

We can see an example of a trophic cascade from Cayuga Lake in upstate New York. The lake in the late 1960s was polluted as a result of sewage input and fertilizer runoff. It was green and smelly and not a particularly nice place

to swim or fish. The food web of Cayuga Lake was fairly simple at the time, but we are going to simplify it even further and represent it as a food chain (Figure 6.9A). You can see that the chain was composed of three trophic levels, primary producers (algae, which dominated and made the lake water green), primary consumers (zooplankton, little microcrustacean animals), and a secondary consumer (small fish called smelt). Resource managers for the lake decided to improve the lake (i.e., make it not green) by adding a tertiary consumer to its food web. The consumer they added was a top predator, a piscivorous (fish-eating) fish, the lake trout *(Salvelinus namaycush)*. What do you think happened? By adding the trout, the population of the smelt decreased, because they now had a predator (Figure 6.9B). With the smelt population reduced, the zooplankton populations were able to increase, which in turn caused a decrease in the algae population. The lake went from being green and dominated by algae to being clear as a result of adding a top predator and having the effect cascade down the food web. This type of dramatic change does not always happen, but what is clear from this example is that food webs are dynamic, interconnected entities where every member is linked to all the other members. Change a food web in one place, and the effects will be felt elsewhere. Sometimes these changes are for the better (from a human perspective), but more often than not, altering food webs has a decidedly negative impact, such as reducing fisheries, as we will see in later chapters.

ENERGY FLOW

So far, all the food chains we have seen have only three or four trophic levels. Why is that? Why is it that there are no predators that feed mainly on bald eagles or black bears or mountain lions? What prevents there from being a fourth- or fifth- or sixth-level consumer in most cases? The answer has to do with energy and the flow of energy between tropic levels. It turns out that, at each step up the trophic ladder, a large amount of energy is lost. Hold on, you might say: didn't some brainy physicist once say that matter and energy are neither created nor destroyed? How is it possible to lose energy, and where does it go?

To find the answer, we are going to turn to a grasshopper and see what happens to the food it eats. For example, by eating grass all day, a really big grasshopper may take in 120 calories. Of those calories, 33 percent (40 calories) is lost to heat and respiration. What does this mean? Calories are a measure of the energy content of food, and heat is one form of energy. When our bodies digest and metabolize food, one of the byproducts is heat. Also, heat is lost to the environment when muscles do work and when animals breathe (remember that air coming out of lungs is warm and holds water). Additionally, another 50 percent of the energy (60 calories) the grasshopper took in as food is lost as feces. Yes, you read that correctly: there are a lot of calories in feces. Bodies are not 100 percent efficient at breaking down food, and there is a lot of energy that essentially passes through an animal because it can not utilize it. Think for a moment about dieting. One old (and not so useful) strategy for losing weight is to eat loads of vegetables such as celery because our bodies do a poor job

of breaking down plant material such as cellulose. Celery is mostly cellulose and water, and so the plant fibers pass right through without adding to our calorie load. Also realize that in many developing nations, many millions of people burn dried cow and horse manure to cook and heat their homes, because there is often a lot of energy left in it.

So what is left after the heat loss and feces? Of the original 120 calories the grasshopper ate, only 17 percent (20 calories) is available to the grasshopper and is made into new grasshopper bits. What this means is that the energy transfer from plant tissue to grasshopper tissue is only 17 percent efficient. In fact, this is true at every single step in a food chain. Only 10 to 20 percent of the energy that is consumed gets assimilated into biomass (tissue) of the new trophic level. To say it another way, energy loss in the transfer between trophic levels is typically between 80 and 90 percent. This simple fact has enormous ecological and conservation consequences. The reason there are rarely fourth- or fifth-level consumers (i.e., predators on top predators) is because there is not enough energy to support them. Think about a predator that only preys on bald eagles and similar large predators. A single super-predator like this would have to hunt over an entire continent in order to find enough bald eagle meat to keep itself alive. There is not enough energy in the bald eagle trophic level to support such a predator.

An exception to the normal shorter food chain lengths occurs in the open waters of the ocean, where food webs are powered by upwelling nutrients and superabundant plankton. Large top predators, such as salmon and whales, can move long distances to find concentrations of food. Thus, a food chain may include more than four levels and could look like this:

phytoplankton → zooplankton → herring → mackerel → salmon → orcas

One reason for this longer food chain is that ocean food webs are very complex, with multiple species at each level.

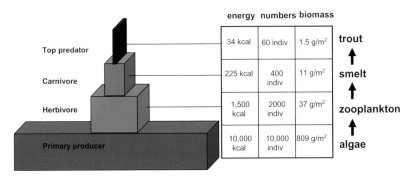

energy	numbers	biomass
34 kcal	60 indiv	1.5 g/m²
225 kcal	400 indiv	11 g/m²
1,500 kcal	2000 indiv	37 g/m²
10,000 kcal	10,000 indiv	809 g/m²

FIGURE 6.10. Ecological relationships among trophic levels, as shown through the use of ecological pyramids, which can be constructed using energy, number of organisms, or biomass. Note the large decreases among all the currencies (i.e., energy, numbers, or biomass) as you go from a lower level of the pyramid to the step above. The Cayuga Lake food chain is included in this figure to make the connection between food chains and pyramids. (Figure by M. P. Marchetti.)

An important conservation consequence of these limits to energy flow is that the higher an individual eats on the food chain, the more energy is needed to support its diet. For example, to generate every 100 calories of beef for your dinner, it takes approximately 1,000 calories of grain. As a result, for the energy it takes to feed one person meat, we could feed 10 people grain. If the world wanted to feed the human population, it would be much less expensive (both ecologically and economically) if we all ate vegetables and grains than if we continued to eat meat. In other words, the meat-focused diet of the western world is costing the planet an order of magnitude more energy and resources than a non-meat diet would, thereby making it harder and less likely to feed the 6.5 billion (and growing) people on the Earth.

VISUALIZING ENERGY FLOW

We can view these energy relationships in a useful manner by constructing what is called an *energy pyramid,* which visually shows the reduction in energy among the trophic levels (Figure 6.10). If you think about it, the amount of energy available at each trophic level also puts a constraint on the actual number of organisms that can be supported on a given amount of energy. In other words, we may have 1,000 grasshoppers eating grass from tens of thousands of grass plants in a field, but this number of insects would feed only 10 shrews. So the amount of energy limits the number of individuals at each trophic level, as we saw above with the impossible idea of a predator surviving by eating bald eagles. To visualize this limitation, we can also construct an ecological pyramid based on the number of individuals at each trophic level or on the amount of biomass (weight of living tissue) (Figure 6.10).

One of the ecological byproducts of this process is a very real constraint on the total number of top predators possible in an area. This becomes an important conservation issue when we do something like reintroduce wolves into Yellowstone National Park. The park is only so big and can therefore only support a limited number of wolves. Wolves above this number will go elsewhere (i.e., outside the park) to find their food, which brings them into conflict with nearby livestock owners. In addition, as we can see in the pyramid for the Cayuga Lake food web (Figure 6.10), such constraints also limit the amount of biomass we can harvest from an ecosystem. If we want to harvest the lake trout from Cayuga lake for food, then we will be limited by the amount of biomass present. We may be able to remove some trout each year, but if

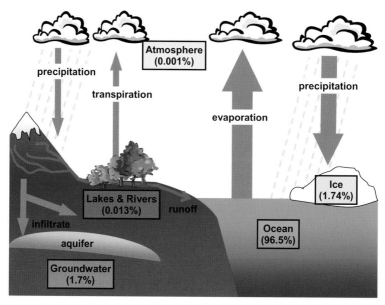

FIGURE 6.11. Hydrologic cycle showing the major places that water is found on the planet. The numbers in the boxes indicate the percentage of the world's total water that is found in the different forms. Note that the vast majority of the water on the planet is salty and is therefore not readily available for human use. (Figure by M. P. Marchetti.)

we remove too many fish, the population may crash, as we saw in the Peruvian anchoveta fishery (Chapter 2), and the system may return to a green lake. Ecosystems are constrained in a very real sense by energy limitations, and as a result, there are literally only so many fish in the sea.

BIOGEOCHEMICAL CYCLES

So far, we have been focusing on the biotic (or living) component of ecosystems, but there is another important player we must now address, the abiotic (or nonliving) side. This is sometimes referred to as nutrient or chemical cycling but is better described as *biogeochemical cycling*. This may sound scary, but in fact what this means is simply following the cycle of nutrients (chemical) through both biological (bio-) and geological (geo-) processes. Many of the largest planet-wide threats to human health and well being, such as air and water pollution, global warming, and ozone holes, deal directly with what happens if one of these

cycles is broken. Here, we discuss three of the biogeochemical cycles: hydrologic, carbon, and nitrogen.

HYDROLOGIC CYCLE

All living beings on the planet need water in some form, so the hydrologic (or water) cycle is vital to life as we know it. In fact, some scientists have suggested that the biggest threat to human welfare is not from dwindling oil supplies, wars, diseases, or even global climate change, but will instead come from shortages of fresh water. With the human population expected to double (from 6.5 billion to over 12 billion) in the relatively near future, the availability of clean water is rapidly becoming a critical issue worldwide. Figure 6.11 shows some of the major features of the hydrologic cycle, including the percentage of the Earth's water in different environments. There are a number of points to notice from this figure. First, it's good to recognize that there is a strong abiotic component to this cycle, as evaporation and precipitation play a large role in moving water around the globe. The oceans are by

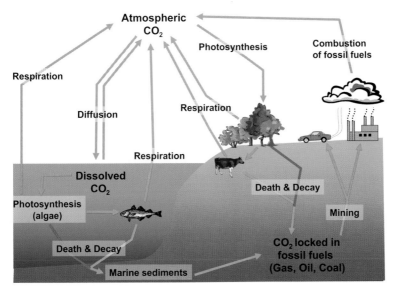

FIGURE 6.12. Carbon cycle showing the major reservoirs of carbon and pathways of carbon movement on the planet. Note the four major pools of carbon, namely oceans, atmosphere, living tissue, and fossil fuels. The burning of fossil fuels has contributed to global warming by greatly increasing the amount of CO_2 in the atmosphere and has the potential to impact it even more in the future. (Figure by M. P. Marchetti.)

far the largest reservoirs of water (96.5 percent), but there is also a significant amount of water locked up in soils and rock called groundwater. It has been estimated that 50 percent of the world's population gets its drinking water from groundwater. This can become an environmental problem in three ways. First, when we remove too much groundwater, it generally takes a very long time to replenish the supply because water does not flow through the soil very quickly. Second, when groundwater gets contaminated through pesticides or toxic effluent, it is almost impossible to clean up because the water is in an underground reservoir. Third, in many areas, salt water from the ocean or other sources creeps in to replace the fresh water being pumped out.

It's also important to recognize the relatively small but significant biological component to the hydrologic cycle. This is mainly due to plant transpiration, which is the process whereby plants give off water vapor. In some parts of the world with large expanses of vegetation, transpiration plays a significant role in producing rainfall, particularly in the tropical rainforests. It has been demonstrated that the clear-cutting of large areas of rainforest has actually caused a decrease in the local rainfall. If we remove a large source of water vapor for the atmosphere, it makes sense that it would affect the amount of rainfall.

CARBON CYCLE

Carbon is also a vital element for life as we know it, and the carbon cycle depicted in Figure 6.12 shows the intimate ties between carbon cycling and living organisms. We can see that there are four major carbon "pools" in the cycle: atmospheric carbon (i.e., CO_2 gas), dissolved oceanic carbon (i.e., dissolved CO_2 gas, as in carbonated beverages), carbon in living tissue, and carbon locked up in fossil fuels (oil, gas, and coal). Yet it is interesting to compare estimates of the amount of carbon in each of these reservoirs. Geochemists suggest there is approximately 0.7 trillion tons of CO_2 in the atmosphere, 1 trillion tons of CO_2 dissolved in the oceans, 1 trillion tons of C in living things, and 5 trillion tons of C in fossil

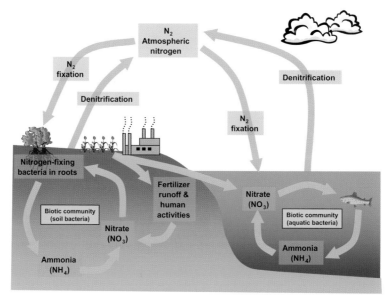

FIGURE 6.13. Nitrogen cycle showing the major pathways and the three major forms of nitrogen: N_2 gas, nitrate, and ammonium. Note the importance of nitrogen-fixing bacteria for the conversion of the largely biologically unavailable N_2 into more useable forms. Because of the relative scarcity of nitrate and ammonium in living systems, excessive use of chemical fertilizer (mainly forms of nitrogen) can lead to unwanted ecological alteration of nearby waterways. (Figure by M. P. Marchetti.)

fuels. With this information, it becomes obvious that burning fossil fuels has the potential to drastically alter the entire planetary carbon cycle and clearly is at the root of the change in global climate we see today. One of the other things to recognize with the carbon cycle is that carbon locked up in plant tissue is hard to liberate because cellulose (which makes up plant cell walls) is difficult for most organisms to break down (remember the desert iguana example). This means that if we wanted to remove CO_2 from the atmosphere and limit global warming, a good way to do it might be to plant more trees while at the same time putting a halt to the clear-cutting of the world's forests.

NITROGEN CYCLE

Nitrogen is a critical component of living beings as it is a building block for molecules such as DNA, proteins, and lipids. Despite this fact, the vast majority of nitrogen on the planet is unavailable for organisms because it exists as atmospheric or elemental nitrogen (Figure 6.13).

A full 78 percent of the atmosphere is made up of nitrogen gas (N_2), which (as you may remember from high school chemistry) involves a triple bond between the two nitrogen atoms. Triple bonds are extremely hard to break, and only a small number of organisms called nitrogen fixers (mostly bacteria) are able to perform this task. Many of these nitrogen-fixing bacteria live in a mutualistic relationship with other organisms, for example in leguminous plants (peas and beans). Because usable nitrogen is so hard to come by, animals and plants readily take in nitrogen whenever they can find it. We get the vast majority of our nitrogen from our food, whereas plants have to scrub the soil for nitrogen, sometimes assisted by mycorrhizal fungi, as we have seen. Fertilizer is largely made of nitrogen (and some phosphorous); when it is applied liberally to agricultural crops, much of it flushes off the field and ends up in streams, rivers, and lakes. Freshwater algae, like other organisms, use nitrogen and often have huge population blooms as a result of agricultural runoff, as we saw in Cayuga Lake. This kind of enrichment of fresh waters with nitrogen is called *cultural eutrophication* and results in a major shift in aquatic organisms, often toward species considered less desirable by humans (e.g., a shift from trout to perch) and toward nuisance blooms of algae.

BROKEN BIOGEOCHEMICAL CYCLES

What happens when these biogeochemical cycles are disrupted? What are the effects, and how does the ecosystem respond? These are important questions and conservation biologists have spent a lot of research energy trying to uncover some of the answers. One of the most famous and vivid examples of how a system responds in the face of drastic change comes from research done in the 1960s in an experimental forest in New Hampshire. The Hubbard Brook Experiment set out to discover what happens to nutrient cycling when all the trees are removed from a forest. The researchers looked at the forests that surround two nearby streams and along one stream cut down all the vegetation in the watershed. They then monitored chemical cycling in the stream outflow and watched over time to see how nutrients changed in the runoff. One of the major findings was that the levels of nitrogen runoff in the clear-cut stream skyrocketed following the logging treatment, suggesting the vital importance of the biological component to biogeochemical cycling. If we break these biogeochemical cycles, as happened at Hubbard Brook, there can be dramatic consequences, the effects of which ripple across all levels of ecological organization, from the insects living in the stream, to the birds feeding on the insects, to the fish and vegetation downstream receiving the nitrogen-enriched water. All organisms living together in a place are connected by a powerful set of biotic and abiotic forces that work together as a whole and produce the bewildering complexity we see in ecological systems.

CONCLUSION

This brings us back to a very basic idea: humans are the biggest ecological force on the planet. We are just beginning to appreciate how our impact on other organisms, as individuals, populations, or as part of communities and ecosystems, can make the world less habitable for humans. Conservation science is all about minimizing human impacts at all levels. This can happen only if we get an understanding of the basic ecology of the world around us and incorporate this knowledge into our actions and activities.

FURTHER READING

Leopold, A. 1990. Sand County almanac. Ballantine Books. New York. *Still in print: read it.*

Walker, B., and D. Salt. 2006. Resilience thinking: sustaining ecosystems and people in a changing world. Island Press. Washington, DC. *An exciting, readable account of our developing understanding of ecosystems and how to live sustainably in them.*

Worster, D. 1994. Nature's economy: a history of ecological ideas. Cambridge University Press. Cambridge, UK. *A classic but still accessible account of the history of ecological thinking.*

7

Biodiversity and Extinction

Conservation science depends a great deal on understanding the nature of biodiversity as well as on understanding the process of extinction. Biodiversity (aka biological diversity), as we will see below, has many facets and encompasses a wide range of ideas. Understanding some of these complexities will allow us to build bridges between ecological and evolutionary science and conservation science (aka conservation biology). In order to do this, we will first define what we mean by biodiversity, and then explore how it can be measured, examine where biodiversity is found, enumerate how many species we are dealing with, and discuss major threats to biodiversity. In order to understand extinction, we will first describe what it means for a species to be extinct, and then look at what scientists know about the process of extinction from studying the fossil record. We will end the chapter with a discussion of the ongoing extinction event.

BIODIVERSITY

Biodiversity has been defined in many ways over the years, but we like the 1989 definition used by the World Wildlife Fund, which defines *biodiversity* as "the wealth of life on earth (all plants, animals, and microorganisms), the genes they contain, and the intricate ecosystems they help build." This is a useful definition for many reasons, one of which is that it is inclusive and really captures the essence of what scientists mean by the term. In addition, it highlights that there are three distinct levels of biodiversity that are important for us to consider. First is the level of the species. *Species biodiversity* essentially considers the wealth of different species on the planet or at some more local geographic level. Species biodiversity includes the obvious organisms such as mammals and birds and trees but also encompasses all other forms of life such as mosses, bacteria, viruses, algae, worms, fungi, and microbes. The total number of species is the way most of us think about biodiversity, but the definition above highlights two additional levels that are important to consider: genetic biodiversity and ecosystem biodiversity.

Genetic biodiversity refers to one of the reoccurring themes of this book, namely natural variation. Variation that is passed on through the genes is the material with which evolution and natural selection work. Clearly, when we think about it, this level of diversity is also extremely important. Without genetic variation, individuals of a species would all be alike and unchanging, and as a result, the species would not be able to adapt when their environment changes. In addition genetic diversity plays a role in *population diversity*. In Chapter 3, we talked about populations with different traits that reflect adaptations to different environments (i.e., ecotypes), such as the flower called yarrow growing at different altitudes. It is variation at the genetic level that allows organisms to adapt to changing environments or to move into new environments. Without natural genetic variation, life as we know it (including all species) would not exist, so including genes in our definition of biodiversity seems not only appropriate but necessary.

Ecosystem diversity refers to geographically defined areas with suites of interacting populations of plants, animals, and microbes. Our definition of ecosystem includes *all living organisms in a place combined with the nonliving characteristics of that place,* meaning that species and ecosystems are intimately tied together. Each ecosystem has its own set of organisms and interactions, which are often defined by humans in terms of the most distinctive organisms and processes: a hardwood forest ecosystem, a temperate lake ecosystem, a Pacific coast rainforest. Implicit in the recognition of ecosystem diversity is that natural systems are more complex and species rich than human dominated systems and that by protecting natural ecosystems we are protecting many organisms and processes about which we know little. A forest, with its myriad species of trees, birds, mammals, insects, and other organisms is more diverse than a neighboring agricultural field with its single dominant plant species and near lack of wildlife.

MEASURING BIODIVERSITY

Inclusive definitions such as the one above are always a good place to start with a giant topic like biodiversity, but on a more practical level,

we need to know how do we measure something like biodiversity. Also, if we can measure it, what do our measures mean, and how do we interpret our results? Let's look at a hypothetical example.

Suppose we have two separate chains of small islands, the High Society Islands and the Low Society Islands. Each archipelago consists of three individual islands. On each island, there is a small number of different species that we represent as letters of the alphabet (Figure 7.1). If you were an ecologist sent out to measure biodiversity, how would you do it? One way would be to count up the total number of different species (i.e., letters) for each island, giving you six values, one for each island, when you were finished. Alternately, you could go to each archipelago and count the total number of different species across the entire chain of islands. In this case, you would end up with two values, one for each archipelago. If you do this for Figure 7.1, what kind of numbers do you get? The numbers we calculated can also be seen in Figure 7.1. The first thing to notice is that all the numbers are not the same. The highest single-island value occurs on the third High Society Island and the lowest on the first of the Low Society Islands. Why does this matter, and how do conservation biologists use this kind of information? These become important questions if you want to protect species biodiversity on the islands. To illustrate the difference here, let's say some non-government organization (NGO) came to you and said it had enough money to protect one island. Which island would you recommend? Likely, you would suggest that the NGO protect the one with the highest number of species in the High Society Islands. But if the same NGO said it had found additional funds and wanted to protect an entire archipelago, you would likely recommend the Low Society Island chain because it contains more total species. Conservation scientists have given names to these two types of biodiversity measurements: *alpha biodiversity* and *gamma biodiversity*. Alpha biodiversity (α) is just the number of species counted in a

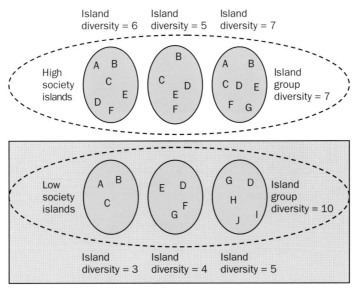

FIGURE 7.1. Species diversity can be measured in a number of different ways. On these two hypothetical island chains (High and Low Society islands), each individual species present on an island is represented by a letter of the alphabet. To calculate diversity (species richness), you simply add up the number of different species (i.e., letters) in a particular area. For example, conservation scientists can measure the species richness of each island, or alternatively they can measure the richness of the entire island chain. Note that when using diversity on an island by island basis, the highest diversity area is in the High Society Islands, but when calculating island chain diversity, the most diverse area is the Low Society Islands. (Figure by M. P. Marchetti.)

local community (or in this case on an island). Gamma biodiversity (γ) measures the number of species across multiple communities (or, in this case, across an island chain).

An additional complication to measuring biodiversity comes if you start looking at genetic biodiversity. For example, on our islands above, you might find that populations of each species on different islands have genetically diverged from other populations to a certain degree and that the degree of genetic difference among populations would likely depend in part on the distance to the nearest other island. If part of your conservation goal was to protect each species' ability to adapt to changing conditions (or to survive a hurricane that might wipe out one island), then you might want to be sure that each island has at least one protected area large enough to maintain each species. But what if you could only protect the population on one island? Which population would you choose?

The biggest one? The one with the highest genetic diversity? The one that is most distinctive genetically? Or just whatever you can get? Unfortunately, the last answer is usually the correct one.

The next difficult question that arises is, How would you protect ecosystem diversity, given limited opportunities? From Figure 7.1, it appears that the third High Society Island has the most species, so likely it has the most intact ecosystems, whereas the third Low Society Island has some species found there and nowhere else (i.e., endemic species), indicating a distinctive ecosystem type. What would your strategy be for protection?

The point of this exercise is to show that the way we measure biodiversity has an impact on how conservation or protection efforts can work in the real world. The choice is determined by the question that is being asked and the problem that needs to be solved. Presumably, in the

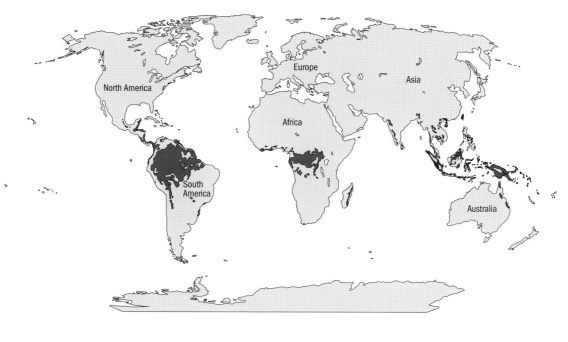

FIGURE 7.2. Map showing the locations of the tropical rainforests of the world. Note that the majority of the rainforests lie near the equator and that all of them lie between 30° N and 30° S latitude. (Reproduced with permission from Rhett Butler at www.mongabay.com.)

ideal world, you would use diversity measures at all three levels to devise the optimal conservation plan.

WHERE IS BIODIVERSITY HIGHEST?

Now that we have some idea regarding the difficulties involved in measuring biodiversity, you may be asking yourself, Where is the greatest amount of biodiversity found on the planet? And is there any pattern to the biodiversity across the globe? In tackling the first question, many people have a pretty good idea where two of the three most diverse biomes (see Chapter 3) are on the planet, but they often are completely oblivious of the third. So, as a little exercise, see if you can guess the three most biologically diverse biomes around the planet.

The first most biologically diverse place nonscientists generally guess is the tropical rainforests. This is indeed a good guess because it has been estimated that, although tropical

rainforests account for only about 7 percent of the Earth's land mass, they house up to 50 percent of all the plants and animals, making them extremely biologically rich. We have already seen why tropical rainforests are found where they are due to climate patterns, and we can see that they indeed cluster around the equator (Figure 7.2).

What you may not appreciate unless you have had the privilege of visiting one is how incredibly complex and abundant life is within a rainforest. In terms of tree species alone, the numbers are staggering. For example, in parts of South American tropical rainforests, you can find upwards of 300 species of trees. Many of us do not know 30 different species of trees, let alone 300, and that is just the tip of the iceberg. Tropical rainforests have a huge diversity of plants partly because there is a three-dimensional characteristic to the forests. The forest is composed of many distinct horizontal layers, including the canopy, the understory, and plants that grow on top of other plants (called

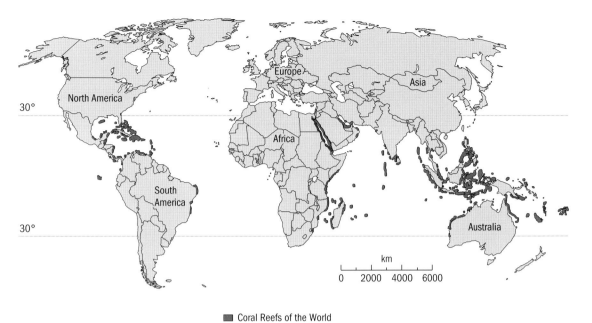

■ Coral Reefs of the World

FIGURE 7.3. Map showing the locations of the coral reefs of the world. Note that the majority of the coral reefs lie at or near the equator and that all of them lie between 30° N and 30° S latitude. Note, many Pacific island coral reef systems (e.g., Hawaii) are not shown on this map. (Reproduced with permission from World Conservation Monitoring Center. 1992. Global biodiversity: status of the Earth's living resources. Chapman and Hall. London.)

epiphytes). When you walk around in a rain-forest, you see a wide variety of plants on the ground, but if you were able to walk around in the canopy, sometimes hundreds of feet above, you would find an almost entirely different set of plants that you would never see from the ground. Also, when visiting a rainforest, it pays to walk slowly and look closely at the small things. Many times, you can find an amazing array of epiphytic plants and fungi growing on top of one another, creating a rich mosaic of diversity.

So far, we have only been talking about plants, but animal diversity is many times more diverse than that of plants. For example, on five small plots in the Amazonian rainforest, scientists collected over 3,000 species of beetles. We are not talking about 3,000 individuals or 3,000 species of all kinds of insects, but 3,000 species of just beetles! And this was only from one tiny sliver of the Amazon. We find similar trends in bees, ants, birds, fish, amphibians, reptiles, and mammals, with new ones being discovered all the time. The tropical rainforests of the world

are amazing showcases of natural wonder and variety.

If we now try to consider the second most biodiverse biome, many people would suggest coral reefs, and again they would be correct. Coral reefs account for only a tiny fraction of the Earth's oceans (0.3 percent), but they contain more than 40 percent of the fish species found on the entire planet. One study from the 1970s found 75 fish species using one coral head that was only a few cubic meters in volume. Coral reefs are the marine counterpart to tropical rain-forests in that they also are found at or near the equator (Figure 7.3) and contain a bewildering three-dimensionality in their habitat structure. In fact, there are many organisms that take advantage of this complex structure and actually live inside the reef itself. You would never see these critters unless you were to chip or break apart the coral, but they are there nonetheless. If you have had the chance to snorkel or dive on a coral reef you, will recall the otherworldly nature of the experience. Coral reef habitats are like none other on the planet in terms of the colors, shapes, and often bizarre adaptations the reef inhabitants have, which allow them to live in the warm tropical waters. It's not just corals and fishes that are found there: coral reefs house incredible numbers of mollusks, shrimps, sea slugs, sea anemones, sea stars, sea cucumbers, crabs, and octopi.

Now we are down to the tricky one, the last of the three most biologically diverse biomes on the planet. We will give you a hint. It's a place that occupies a very large portion of the globe, but it's an entire biome that you and everyone you know will likely never visit. It also may be the second most biologically diverse biome on the planet, at least in its total number of species. The answer is the deep ocean floor, below about 100 meters depth. The deep ocean floor covers an enormous percentage of the planet and has been almost completely unexplored until very recently, thanks to advances in deep sea technology. What do you find down there in the dark cold waters of the deep ocean? Marine biologists are finding an almost unbelievable diversity of marine organisms such as clams, mussels, worms, sea stars, tunicates, protozoa, crabs, nematodes, and shrimps. A gigantic 10 year international oceanographic survey called the Census for Marine Life, started in 2000, has been surveying parts of all of the world's oceans. One of the survey techniques is to drop big marine dredges (i.e., bottom scoopers) to the deep ocean floor. The dredge samples have been bringing up somewhere near 2,000 new species a year. The biologists sample one location and find new species never seen before and sample again a relatively short distance away, only to find entirely new species. Some estimates coming out of this marine survey (set to be completed in 2010) suggest that, in the deep ocean sediments alone, there are likely upwards of 750,000 species, most of them undescribed.

LATITUDINAL TRENDS

One of the other major trends in biodiversity you may have already noticed from this and previous chapters is that we seem to find more species near the equator and fewer near the poles. We can see these latitudinal trends by looking at maps showing the distribution of certain groups of organisms (Figures 7.4–7.6). Figure 7.4 shows the worldwide distribution of amphibian species. There are a few important things to notice from this map. First is that amphibian species are clearly not distributed evenly across the planet; there is a much higher concentration near the tropics than near the poles. Additionally, it's interesting to note that, even within the same latitudinal band, there is an uneven distribution. For example, eastern North America has many more amphibian species than western North America has at the same latitude. Yet this does not always hold. If we were to look at a map showing the distribution of mammal species in North America, we would see the opposite trend, with more species in the west than in the east.

Biodiversity is clearly not evenly distributed. We see similar trends in bird diversity across North America, although because many birds migrate, the patterns change dramatically between winter and summer (Figure 7.5). When we look at a map of vascular plant diversity around the world (Figure 7.6), we see again the general trend of more diversity near the equator, but we also see other seemingly isolated areas of high diversity (e.g., the southern tip of Africa or the southwestern portion of Australia). These areas are sometimes called biodiversity hotspots and often differ in location depending on the group of organisms you look at. For example, the table below (Table 7.1) shows data for some of the world's biodiversity hotspots and the number of endemic taxa found in each area (note that an *endemic* organism is one found only in one region or locality).

One of the interesting features from this table is the fact that some taxonomic groups are extremely diverse in some areas but not others. For example, the endemic amphibian diversity in coastal Brazil is extremely high, as is the reptile diversity on the island of Madagascar. All of this shows us that, although we can find trends in biodiversity, the world's

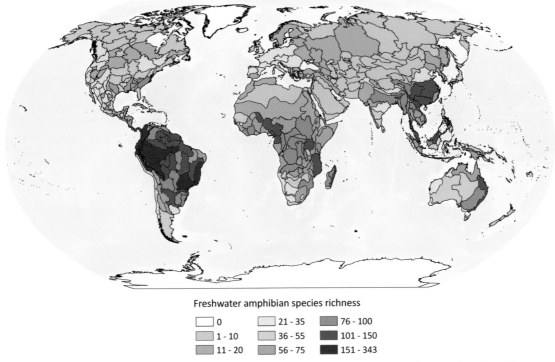

Freshwater amphibian species richness

☐ 0	☐ 21 - 35	■ 76 - 100
☐ 1 - 10	☐ 36 - 55	■ 101 - 150
☐ 11 - 20	☐ 56 - 75	■ 151 - 343

FIGURE 7.4. Map showing the global distribution of amphibian species by watershed. Note the large numbers of amphibian species near the equator, but also note the non-equatorial hotspots, including the southeastern portion of the United States. (Reproduced with permission from IUCN Species Program and Conservation International Center for Applied Biodiversity Science.)

organisms are not spread out evenly over the planet's surface.

NUMBER OF SPECIES

So far, we have skirted around two of the central questions about biodiversity, namely how many species there are and which taxonomic groups have the most diversity. As we will see below, these are not easy questions to answer, and we will have to rely on some rough calculations in order to derive reasonable estimates. One of the first places to start this inquiry is to take a look at how many species scientists have described. When a new species is described, the description results from taxonomic experts looking at a collection of specimens and comparing the characteristics with those of all similar known organisms. If the specimens are found to be different enough (see Chapter 3 for a discussion on the biological species concept), then the group

to which they belong is given a two-part Latin name (genus and species) and is considered a new species. If we examine only those species that have been described by scientists in this way, we come up with a total number of about two million species, which sounds like quite a few. Figure 7.7 details the taxonomic groups for the majority of these described species. In this, we can see that all the higher animals (birds, reptiles, amphibians, fish, and mammals) only make up about one-eighth of the total number of species described by science and that by far the largest and most biologically diverse group is the insects, which account for fully half the total number.

You may notice a tilde before all the numbers given in Figure 7.7. This indicates that the numbers are approximate values; we don't have exact figures. Partly this is a result of new species being described all the time. For example, expeditions to poorly documented parts of the

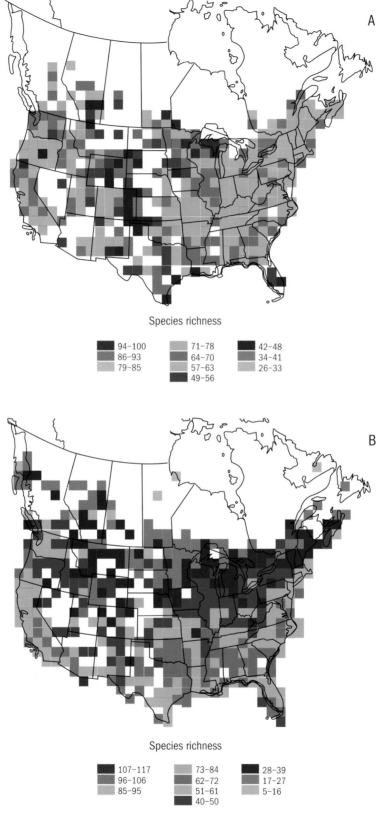

Species richness

94–100	71–78	42–48
86–93	64–70	34–41
79–85	57–63	26–33
	49–56	

Species richness

107–117	73–84	28–39
96–106	62–72	17–27
85–95	51–61	5–16
	40–50	

FIGURE 7.5. Two maps showing bird diversity across North America in two different seasons, summer (A) and winter (B). The change in diversity is the result of many species of birds migrating south to avoid the short, cold days of winter in the northern hemisphere. (Reproduced with permission from Hurlbert, A. H., and J. P. Haskell. 2003. The effect of energy and seasonality on avian species richness and community composition. The American Naturalist 161:83–97.)

GLOBAL BIODIVERSITY: SPECIES NUMBERS OF VASCULAR PLANTS

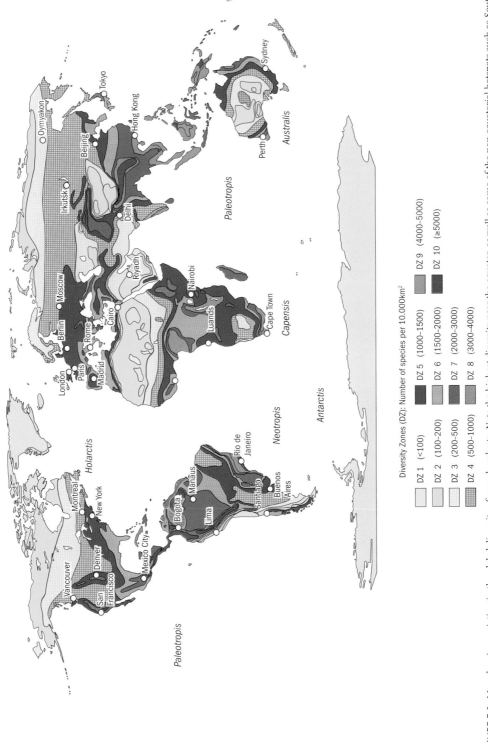

Diversity Zones (DZ): Number of species per 10.000km²

DZ 1 (<100)
DZ 2 (100–200)
DZ 3 (200–500)
DZ 4 (500–1000)
DZ 5 (1000–1500)
DZ 6 (1500–2000)
DZ 7 (2000–3000)
DZ 8 (3000–4000)
DZ 9 (4000–5000)
DZ 10 (≥5000)

FIGURE 7.6. Map showing variation in the global diversity of vascular plants. Note the highest diversity near the equator, as well as some of the non-equatorial hotspots such as South Africa and southwestern Australia. (Reproduced with permission from Barthlott W., et al. 2005. Nees Institute for Biodiversity of Plants, University of Bonn.)

TABLE 7.1

Numbers of Taxa in Some Diversity Hotspots

REGION	MAMMALS	REPTILES	AMPHIBIANS
Coastal Brazil	40	92	168
Tumbes-Chocó (South America)	60	63	210
Philippines	98	120	41
Borneo	42	69	47
Southwestern Australia	10	25	22
Madagascar	86	234	142
South Africa	16	43	23
California	15	25	7
New Caledonia	2	21	0
Indian Ghats	7	91	84

world are still turning up many new species of vertebrates. Biologist James Patton, visiting the Andes Mountains in South America in 1995, discovered, in two weeks time, six new species of mammals: four rodents, a shrew, and a marsupial. Recent biodiversity exploration in the Foja Mountains of Indonesia discovered 20 new frog species, including a tiny frog less than a half-inch long, four new butterfly species, and at least five new species of palm plants.

Clearly, all the species on the planet have not been described by science, so our first cut at a total number of species on the planet is going to be low. But the question is, How low is our estimate of two million species? How many more species are hiding out there (remember the 750,000 estimate for deep sea species alone!)? This is where things start to get tricky. Clearly, we do not know the absolute number, so scientists have had to develop ways to estimate the total number. Some of these estimates come from studies like the one already described where 3,000 species of beetles were found in a few small square meter blocks of rainforest. If we take this number and scale it up to the entire rainforest, we can then have a ballpark figure for the total number of beetle species in the Amazon. And if we assume that the same pattern holds across all rainforests, then we can scale it for the total acreage of rainforest

How many species exist?
Described species ~ 1.9 million

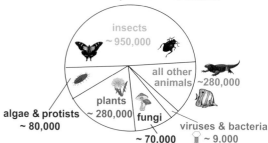

FIGURE 7.7. The approximate numbers of described species worldwide. Note the large number of described insects. (Figure by M. P. Marchetti, with data taken from World Conservation Monitoring Center. 1992. Global biodiversity: status of the Earth's living resources. Chapman and Hall. London.)

on the planet. Similarly, scientists have found large numbers of other taxa in small areas; for example, over 200 species of nematode (tiny soil-dwelling round worms) have been found in just a few grams of earth. Some ecologists have taken all these estimates for different taxonomic groups and put them together. Figure 7.8 puts this all together and shows both the total number of described taxa and the estimated numbers for many taxonomic groups.

We can see some very interesting results from this exercise. For example, in some groups such as vertebrates and plants, we have done a pretty good job describing a large

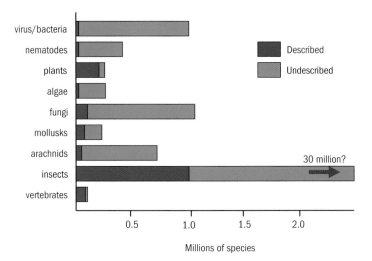

FIGURE 7.8. Estimated total planetary biodiversity showing the approximate number of described species (dark bars) and the estimates of the number of undescribed species (light bars). Note how certain taxonomic groups (e.g., vertebrates and plants) have been fairly well researched, whereas other groups (e.g., insects, fungi) are very poorly known. (Figure by M. P. Marchetti, with data taken from World Conservation Monitoring Center. 1992. Global biodiversity: status of the Earth's living resources. Chapman and Hall. London.)

fraction of the diversity, but for other groups like nematodes, fungi, and arachnids (spiders and mites), we have done a very poor job. The other thing to notice from these estimates is that when we try to estimate the diversity of insects we find that there could easily be 30 million species of insects alone, which is a mind-boggling number. If we put all this information together, the best estimates for the total number of species range from five to 50 million species, either of which is an enormous number.

One question might occur to you when looking at this gigantic estimate of the number of species is, Why is there such a large range? Why can't we get a better handle on this number? If we told you that the price of a new car you wanted to buy was somewhere between $5,000 and $50,000, you might be a little put off. Are the scientists doing this kind of work just bad at their jobs? It turns out that there are four major reasons why we have such an imprecise estimate for planet-wide biodiversity. The first reason has to do with how little the planet's life has been examined. We have a pretty good handle on the number of plants and animals in places like New York, Massachusetts, Nevada, Great Britain, Japan, and Italy. There are a lot of people who live in these places, so they are well explored. They are also the major

centers for ecological and taxonomic work, so they have been the subject of many scientific research projects. This is in contrast to places such as the rainforests of Papua New Guinea or the Amazon or Bhutan or Myanmar. These places are more remote and have only been under close scientific scrutiny for the last 50–100 years.

A second reason has to do with the physical size of many of the most diverse taxonomic groups. Insects, spiders, nematodes, fungi, and bacteria are all tiny and often hard to see, which makes them both hard to collect and hard to identify. In addition, many of the most taxonomically diverse groups are found in places that are difficult for humans to get to. For example, some species of rainforest amphibians live 30 or more meters up in the canopy of the forest and never come down to the ground at any point in their lifecycle! It is very difficult for scientists to access or survey the canopy of a rainforest. In addition, it's very hard to collect deep oceanic organisms; you need either a multimillion dollar submarine or you have to blindly drop a dredge hundreds of meters down in the ocean. This has been described as similar to trying to catch a butterfly in Manhattan from New Jersey using a net mounted on a very long pole. This is pretty difficult if what you are trying to catch can swim away.

The final reason why we are so imprecise with our estimates of biodiversity lies with the fact that the skills necessary to compare and identify new species are esoteric and very difficult to master. For example, there are not very many nematode or fungus taxonomists out there in the world. It's very hard work, often tedious and repetitive, and it takes an enormous amount of training and practice. In addition, the number of taxonomists is not growing very rapidly because fewer and fewer graduate students are choosing to pursue this line of study. This is in stark contrast to Charles Darwin's time, when being a naturalist or natural philosopher (as early ecologists were known) entailed learning how to identify and classify living organisms. These days, there is little funding or professional glory involved in describing ant species, although some of the most famous modern biologists (e.g., E.O. Wilson) are gifted taxonomists who have focused on identifying obscure species.

Given the shortage of taxonomists, one way to get a handle on species diversity that is becoming increasingly popular is to run DNA tissue samples through automatic processors looking for patterns. A distinctive DNA pattern is assumed to represent a new species. This kind of DNA record will not allow sight identification of a new species, but it can be used to quantify the number of species of various taxonomic groups (e.g., beetles) in a region without going through the tedious process of a formal species description. A recent example of the usefulness of this technique occurred when marine ecologists examined samples of sea water, taken from on top of coral reefs, in a search for distinctive DNA types. Their fascinating results showed not only that there was a huge number of tiny undescribed species of plankton (picoplankton) but also that pristine coral reefs had significantly more species of picoplankton (i.e., plankton with unique DNA fragments) than did reefs impacted by human use and abuse. It is important to note that marine plankton are incredibly abundant in the oceans as well as incredibly important for cycling nutrients such as nitrogen and carbon for the entire planet. So identifying patterns in marine biodiversity is vital to setting conservation priorities in the oceans.

DIVERSITY OF VERTEBRATES

Vertebrates are large, conspicuous animals that we arrogantly assume rule the Earth because we are vertebrates and possess all the seemingly superior traits of that group. The reality, of course, is that invertebrates, such as nematodes, and microorganisms, such as bacteria, dominate the Earth in terms of numbers, biomass, and effects on ecosystem processes. Nevertheless, we provide a brief introduction to vertebrate diversity here because we often emphasize vertebrates in this book and because they are the best documented of major animal groups. Despite their good documentation, many new vertebrate species are described every year.

FISHES Fishes constitute the oldest group of vertebrates, with origins dating back some 400 million years ago. The 27,000 species of extant (living) fish are usually subdivided into three groups based upon anatomical characteristics. The jawless vertebrates (lampreys and hagfishes), with many characteristics of the ancestral vertebrates, include about 80 species. The Chondrichthyes, or cartilaginous fishes, are the sharks, rays, skates, and relatives and are represented by about 850 species. The Osteichthyes, or bony fishes, are the familiar perch, catfish, bass, trout, and relatives and are by far the most diverse group with more than 26,000 species. Although the oceans cover 70 percent of the Earth's surface and contain 97 percent of the Earth's water, only 58 percent of the fishes are marine. The rest live in fresh water, in the many isolated lakes and streams on our continents. The smallest vertebrates in the world are fish, with several species being only 6–8 millimeters long and weighing a fraction of a gram.

AMPHIBIANS Amphibians are dependent upon water for reproduction but are otherwise quasi-terrestrial animals that live part of their lives on land. They descended from fishes about 350 million years ago and today are represented by about 9,800 species such as the familiar frogs, toads, and salamanders. To see newts and salamanders and hear tree frogs in their annual "rite of spring," visit your local pond or temporary stream in early spring.

REPTILES   Reptiles, which evolved from amphibian ancestors, were the first truly terrestrial vertebrates. Some, such as marine tortoises, have since "reinvaded" the sea. Once, reptiles were the dominant vertebrates on Earth (210–65 million years ago, known as the "Age of the Dinosaurs"); there are approximately 6,300 species of reptiles today. Reptiles are represented by such animals as snakes, lizards, and turtles.

BIRDS   Birds arose from reptilian ancestors about 150 million years ago. Birds are now represented by approximately 9,100 species and are well adapted for their largely aerial existence, although a few forms (ostriches, rheas, and several others) have lost the power of flight and are now completely terrestrial. Others, such as penguins, have decided to mimic fish and have become aquatic.

MAMMALS   Mammals, the hairy group of animals to which we humans belong, are thought to have arisen over 200 million years ago. Early mammalian ancestors spent most of their early history skulking in the bushes, avoiding dinosaurs. Mammals include some very small specimens (shrews and some bats, which weigh less than 4 g), as well as some of the largest vertebrates (blue whale, which weighs over 160,000 kg). The mammals include aerial (bats), marine (whales, dolphins, porpoises), and terrestrial forms. There are approximately 4,700 described species.

BIODIVERSITY THREATS

Now that we have a better handle on what biodiversity is, where it is found, and how much of it there is, we need to describe some of the current threats to biodiversity. Historically, the distribution of life on Earth was largely determined by interactions with climate, geography, and other organisms. Now humans have become the dominant "force" that all creatures have to contend with. The threats from this domination fall into five broad categories that we will examine below. Much of the rest of this book and a large portion of the science of conservation are concerned with describing, halting, or trying to repair the damage caused by human threats to planet-wide biodiversity. Here, we will briefly introduce each of these threats, but many of them will be examined in greater detail in later chapters.

HUMAN POPULATION GROWTH

The human population seems to defy what we have learned about natural populations: that they can not exponentially increase forever and that there is an environmental carrying capacity that will slow and eventually limit population growth. Over the last few hundred years, the human population has grown enormously and is currently above 6.5 billion individuals (Figure 5.14). Each and every one of these people, including you, wants food to eat and a roof over his or her head. From the animals we consume, to the cars we drive, to the wood for our houses, each and every one of us uses products derived from the world's animals and plants. As a result, natural resources on the planet, including all levels of biodiversity, are under siege in ways we can hardly imagine. We all want to the same basic needs to be met, and we therefore continue to consume plants and animals at an alarming rate. The greatest threat to the planet's biodiversity comes from our very being—from the fact that we need to consume other organisms in order to live. We humans are the ultimate predators and competitors with other organisms. Physics and ecology both tell us that eventually there must be some limit to population growth. The question is, Will that limit be reached before the human population eliminates a large fraction of the Earth's species and precipitates extinction of a magnitude not seen since the age of dinosaurs? Below, we briefly discuss the major means by which we humans are altering the planet and reducing biodiversity.

HABITAT DESTRUCTION   Clearly, if the human population keeps growing and the demand for goods and services increases along with that growth, then more and more habitat is going to be destroyed. Destruction is happening across the globe in virtually every habitat conceivable, from the cutting of rainforests, to the draining of wetlands, to the bleaching of coral reefs, to the ripping up of deserts to build houses and install golf courses, and even to the melting of ice on the continent of Antarctica. In each and every place human beings go, there is an alteration of natural habitat, which is a major cause of biodiversity loss. The process is exacerbated wherever the human population density is high. Take away or alter the habitat, and there is little chance that the local non-human inhabitants will be able to survive the change. Most monkeys, bears, frogs, wildflowers, and mushrooms require relatively unchanged habitats in which to live, where natural ecosystem processes are still working in their favor. When human beings clear or destroy wild lands, biodiversity is lost, clear and simple.

A feature that generally accompanies habitat destruction is habitat fragmentation. When a once continuous habitat is broken into isolated fragments, many of the organisms that depend on the habitat are lost because the fragments are too small to support a viable population. Fragmentation in some situations is as bad as outright loss.

ALIEN SPECIES As we will explore in great detail in Chapter 11, the movement of species by humans around the planet has caused gigantic changes in the natural flora and fauna. Many significant ecological problems have occurred, and many species are threatened as a result. Invasions by alien species have been identified as the second leading cause of biodiversity decline behind habitat destruction, and the two usually go hand in hand. One of the hardest parts of this is that sometimes we don't even know we are moving species from place to place until it is too late and the damage has been done.

CLIMATE CHANGE Global warming is the result of large-scale pollution of the atmosphere with carbon dioxide and other compounds spewing from our cars, factories, and everything else

that burns fossil fuels. Worldwide, habitats for wild organisms are changing at a rapid pace. As climate warms, precipitation patterns change, ice caps melt, sea level rises, and ocean currents change. Many forms of life lack the capacity to adapt to this rapid change or to move to new areas. The declaration of the polar bear as an endangered species by the U.S. Government in 2008, because of the loss of polar ice, is a symbol of what is happening worldwide. The polar bear may be doomed to survive only in zoos, and even this fate is not possible for most species displaced by climate change.

POLLUTION The 1960s saw a cultural change in the United States, as more and more people began to see the effects of pollution on themselves and their environment. Many good pieces of legislation were enacted in order to protect resources such as air and water, as we will see. Yet for all the advances, the problems of pollution have not been solved, only slowed. Indeed, many of our current pollution problems stem from technological fixes to older problems. As one toxic product becomes banned when severe environmental impacts are found, a hundred more rise up to take its place (e.g., DDT vs. other pesticides), many more toxic than the original. Many of these pollutants are highly volatile and are carried in the atmosphere to places far from their source. Thus, marine communities, tropical forests, Arctic ecosystems, freshwater lakes and rivers, and all the organisms that live in them are threatened by a growing list of pollutants and toxic chemicals.

EXTINCTION

All of the problems outlined above are considered threats to biodiversity because they all directly or indirectly contribute to the extinction of life forms on the planet. But what does it mean to be extinct, and how do we know that the current extinction crisis is different from those that occurred in the past? We explore these questions and some others below, but first it seems necessary to define what we mean by the word *extinction*. When we say that a species has gone extinct, scientists mean that there are no more individuals of that species alive to reproduce: the species has ceased to exist as a life form on the planet. This sounds pretty grim and final, and it is. Extinction is a finality, a terminus, a period at the end of a species' sentence. When a species is extinct, it is gone forever and can never be revived. The world will never again hear the call of the passenger pigeon, nor ever watch a dodo bird or catch a thicktailed chub. These species and thousands of others exist only in our memories, as words on a page, or as photographs in a book.

Extinction is an evolutionary end point as well because it carries the loss of genetic history. If you think about it, each species alive today is essentially a living legacy of millions of years of evolution. The genetic makeup inside every living organism that directs how it acts and moves and reproduces is the result of the forces of natural selection playing out through time. When a species is extinct, this story of its history and the genes that tell that story are both gone for good.

In reality, it is very hard to concretely document that a species is completely gone. How can you tell that there is not one single ivory-billed woodpecker *(Campephilus principalis)* left in the wild? Clearly, it is not physically possible to check every habitat and every potential habitat in the world. This is especially true if you are dealing with something tiny and cryptic, such as the golden toad *(Bufo periglenes)*, a small tropical species of toad last seen in rainforests outside of Monteverde, Costa Rica, in 1989. No individuals of this amazing species have been seen for over 20 years, yet it is still possible that we have not been looking in the correct place. This toad spends most of the year underground in the rainforest and was always spotty in its distribution, so there is still some faint hope that a population will eventually be found. Scientists sometimes say that species like the golden toad are *presumed extinct* because of this difficulty. Once in a great, while a species that was declared extinct is found again in the wild, as was recently the case with the ivory-billed

woodpecker mentioned above. In this situation, the species had been thought extinct for decades before teams of ornithologists supposedly found a few individuals in wetlands in both Arkansas and Florida. Currently, it is believed (but not confirmed) that a tiny population of these birds may be still clinging to existence.

Ecologists also often describe species as being *locally or regionally extirpated*. In these cases, a local population has been lost, but the species still persists elsewhere on the planet. For example, a California endemic fish species, the Sacramento perch *(Archoplites interruptus)*, once was found widely throughout the state's great Central Valley, in streams and rivers. For a host of reasons, including competition with non-native fish and massive alterations to its habitat, the species was driven to extinction in its native range. Luckily, the fish possesses some physiological characteristics that made it a good choice to stock for recreational fishing in alkaline lakes in other places. As a result, the species was introduced into a number of reservoirs in California and Nevada. So even though it is extirpated from its native habitat, the species is not globally extinct, although recent work by ecologists from University of California, Davis, suggests that the species is still in serious jeopardy.

PAST EXTINCTION EVENTS

Although our society officially regards the extinction of a species as a bad thing, throughout the evolutionary history of life on Earth (approximately 3.5 billion years), there clearly have been many millions of extinctions. In fact, it has been estimated that perhaps 99 percent of the species that have ever existed on the planet have gone extinct. So why is the current extinction of a few species such a worry for ecologists and conservation biologists? In order to answer this, we have to turn again to the fossil record and look a little bit at the history of life on the planet.

When discussing the process of macroevolution in Chapter 2, we spent a little time dealing

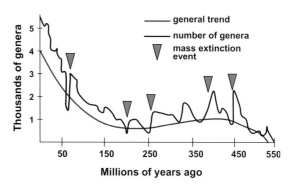

FIGURE 7.9. Trends through geologic time in the estimated number of marine animal genera on the planet, based on the fossil record. Note the general increase in the number over time as well as the sharp decreases during the five mass extinction events. (Reproduced with permission from Rohde, R. A., and R. A. Muller. 2005. Cycles in fossil diversity. Nature 434:208–210.)

with the vicissitudes of the fossil record and learned that, although it is far from a compete accounting of life on the planet, it is in fact a useful tool to understand our evolutionary past. We can therefore use the fossil record to see what it has to say regarding past extinctions.

The fossil record shows a general trend of increasing numbers of living species (or in this case genera—remember that fossil species are hard to identify) through time, as shown in Figure 7.9. We can also see five periods marked on the graph when there were large, rapid losses of planetary biodiversity. These are known as mass extinction events and are defined as the disappearance of at least 50 percent of the species on the planet as recorded by the fossil record. The most well known of these is the most recent, the Cretaceous–Tertiary event 65 million years ago, which resulted in the extinction of the dinosaurs. This event is believed to have been caused by massive environmental collapse precipitated by the impact of a large asteroid.

These past mass extinction events altered the very trajectory of life on the planet and at times completely reshuffled the evolutionary deck, allowing certain taxa to dominate for periods of time (i.e., dinosaurs and now mammals). Yet these events are not the norm; they were likely precipitated by extraordinary circumstances.

We can look to the fossil record and estimate what the normal or background rate of extinction of species might have been outside of the periods of mass extinction. When we do this, scientists find that background rates vary for different taxa, but an average rate seems to be one extinction per every million species years. This sounds confusing, but essentially it means that if there were a million species, the background rate of extinction would be one species a year going extinct. So, if the global estimates of biodiversity are correct at the high end (50 million species), then we would expect an average of 50 species to go extinct each year, out of the 50 million. This may sound like a lot, but during the mass extinction events of the past, more than half (50 percent) of the species on the planet disappeared very rapidly. The background rate described above is a loss of 0.000001 percent, a much, much smaller percentage. Think about the background rate this way: if you had 50 million pennies in a pile, and I took away 50 of them a year when you were not looking, then you would not likely be able to tell that any were missing for thousands of years. In other words, species do tend to go extinct at a natural rate, but it is a very, very slow rate and would be all but imperceptible to our eyes.

THE SIXTH MASS EXTINCTION

The vast majority of biologists are gravely concerned by the current rate of biodiversity loss and believe our current population growth and our associated practices are creating the sixth mass extinction in the history of the Earth. That is correct: we just said that *we are in the midst of the sixth mass extinction event on the planet right now.* Why do we think this, and what sorts of evidence do we use to arrive at such a dire prediction? With the advent of satellite imagery and sophisticated mapping tools (think Google Earth but incredibly more powerful), scientists can actually watch the destruction of the world's tropical rainforests week by week. Using these tools, it has been estimated that over 80,000 square miles of forest are cut down each year. Living in, on, under, and around those forest trees are thousands of species, which are lost, driven to extinction when the forest is cut down. Using evidence such as this, the current rate of extinction is estimated to be at or above 100,000 species lost every year, a full 2,000 times greater than the normal background rate of extinction.

What does 2,000 times greater actually mean? To get a better sense of how fast this is actually occurring, we can do a tiny thought experiment. If you were driving your car at a speed of 100 kilometers per hour (approximately 60 miles per hour), in 60 minutes' time, you would travel 100 kilometers. If your rate was 2,000 times as fast, then you would be moving at 200,000 kilometers per hour, and in 60 minutes of driving, you would make it around the Earth five times! Given that the current rate of extinction is 2,000 times greater than normal, it is estimated that over one-third of all mammal species on the planet will be extinct in the next 30 years! By the end of the century (year 2100), it's likely that upwards of 50 percent of all the species on the planet will be extinct. Think about it: this is half of all the different forms of life disappearing in a few hundred years. This is an incredible and very frightening number of species to be wiped out in a very short period of time. Species losses on this scale are hard to comprehend and make many of us scientists extremely scared for the future of the planet and very, very sad.

The sixth mass extinction is upon us and is so clearly the result of human activity that at times it is actually hard to see. The majority of scientists rate biodiversity loss as a more serious environmental problem than the depletion of the ozone layer, global warming, or pollution, although all of these are clearly tied together. All of humanity, not just scientists, needs to be worried about this and must act now to address this sixth major extinction event.

CONCLUSION

Biodiversity, the incredible wealth of life on the planet and the evolutionary legacy of hundreds of millions of years, is under siege and is disappearing so fast that we often do not know what we are missing until it is too late. Our understanding of what constitutes global biodiversity is still in its infancy and is in a race against time. Yet there is a growing realization that protection of biodiversity has enormous benefits to humans, as we will see in later chapters. We will also see that we need to find ways to support, rather than eliminate, the ecosystems and species that provide us benefits. The protection of the diversity of life on Earth requires both a grand global strategy and intensely local and regional efforts to support day-to-day conservation measures. The majority of scientists believe the crisis could be averted by a stronger stance by policymakers and governments and by individuals making changes in their daily lives. This is at the heart of what conservation science is all about: life, diversity, evolution, and protection.

FURTHER READING

Carwardine, M., and S. Fry. 2009. Last chance to see. Collins Publishing. New York. *This is a very accessible introduction to extinction. Douglas Adams (the author of the* Hitchhiker's Guide to the Galaxy) *and the authors travel to a few remote places in order to see a number of animals on the brink of extinction.*

Quamman, D. 1996. Song of the dodo: island biogeography in an age of extinction. Scribner. New York. *An eloquent exploration of the world's distant places and the obscure species they support, and the tragedy of the loss of dodos and similar species.*

Wilcove, D. S. 1999. The condor's shadow: the loss and recovery of wildlife in America. W. H. Freeman. New York. *A sad yet hopeful book on endangered species.*

Wilson, E. O., and F. M. Peter. 1988. Biodiversity. National Academy Press. Washington, DC. *This is, in some ways, the most accessible book on biodiversity out there and is highly worth reading.*

8

Value, Economics, and the
Tragedy of the Commons

At this point, we have explored some details of the planet's biodiversity and highlighted the biodiversity and extinction crisis caused by human activities. Key questions still remain. For example, why should we care about these irreversible changes to our planet? Isn't improving the global economy worth the sacrifice of a few obscure trees, fish, and spiders? In fact, there are many sound reasons for keeping the planet's biodiversity intact because obscure trees, fish, and spiders make the Earth a better place for humans. Conversely, loss of biodiversity ultimately will make our planet a worse place for humans. The reasons for this pervade our lives and include issues of health, religious beliefs, and economic well being. In this chapter, we briefly explore the complex subject of values in relation to conservation. We discuss how economic thought and reasoning in particular can be a double-edged sword for conservation efforts. This forces us to confront one of the most difficult problems facing biological conservation, the issue of common access resources. Following this, we briefly explore the field of ecological economics to provide insights on how to achieve truly sustainable economic systems.

VALUING BIODIVERSITY

If you were to ask 100 people on the street why they should care about wildlife, as well as more obscure organisms, you would get close to a hundred different answers. Caring about or valuing a thing is a fundamental human endeavor, and people's value systems are often idiosyncratic, shaped by their education, life experience, and cultural surroundings. The hundred different responses would each take a different approach to answer the same question we have asked: why does biodiversity matter? Sociologists and philosophers have looked at people's answers and have grouped them into different categories. We want to present five general sets of reasons for you, knowing full well that this is not an exhaustive list but is instead an attempt to organize an inherently complex set of ideas.*

---

* Many philosophers, scientists, and poets have written eloquently about valuing nature. If you want to delve into this area in more detail, read works by Gary Snyder (*Practice of the Wild*, 1990), Holmes Rolston (*Conserving Natural Value*, 1994), Bryan G. Norton (*Why Preserve Natural Variety?*, 1987), Edward O. Wilson (*The Biophilia Hypothesis*, 1995; *Consilience: The Unity of Knowledge*, 1998), and David W. Orr (*Earth in Mind*, 1994, and more recent essays).

DIRECT ECONOMIC VALUE

To see why someone would place a value on biodiversity, perhaps it is easiest to begin the discussion in terms of money. Monetary or economic value is something that we intuitively seem to grasp from an early age. Questions involving how much a thing costs and how much someone is willing to pay for it are convenient ways for us to gauge the value of something. Biodiversity can be approached in this same manner. How much, in terms of resources, can be extracted from a particular organism? What is biodiversity's monetary value? In other words what is the utility of biodiversity for humans in terms of cold hard cash? This is a very anthropocentric way of approaching value, but it is clearly a way that can be shared and compared among individuals, communities, and countries. It is also generally understandable to most people and allows us a method for accounting.

We can consider direct economic valuations in two different categories: consumptive use value and productive use value. *Consumptive use value* is essentially the value of animal and plant products that are used directly by people. These are natural products that are harvested or collected or taken directly from the natural world without ever being exchanged for money, such as wild mushrooms or wild game caught for subsistence or sport (Figure 8.1). For example, let's say you live on a 30 acre parcel of forested land that has a trout stream running through it. And let's say that throughout the entire year you use wood from the forest for cooking and eat trout you catch from the stream, garnished with wild mushrooms. You would clearly place a large value on the products of biodiversity (wood, fish, and mushrooms), but you did not exchange money for the products that come from nature. This is still an economic enterprise because we can assign a monetary value to the products based on what others would have to pay for the same products or based on what your replacement costs for wood, fish, and mushrooms would be if you moved off your 30 acres. In other words, people value biodiversity

FIGURE 8.1. Delicious morel mushrooms *(Morchella esculenta)* collected in the wild for eating at home are one of many examples of biodiversity products that we can value for their consumptive use. (Photo by M. P. Marchetti.)

because they benefit from it in this economic manner.

Generally, this direct consumptive use is part of more traditional or aboriginal societies, and as a result, a majority of people throughout the world still value and interact with biodiversity in this manner. For those of us living in cities, we are often hard pressed to find anything in our day to day life that we acquired directly from nature without paying for it, unless we occasionally hunt for our food or raise some of our vegetables in a garden.

Consumptive use poses an interesting problem for economists interested in estimating the monetary value of wildlife and biodiversity. Because consumptive use does not involve the exchange of money, there is no record of the amount or value of the products used, so this type of economic activity essentially goes unnoticed by economists. Consumptive use value is not considered in calculating a country's gross national product (GNP). In the past, a developing nation may have derived a large fraction of its total economic worth (in terms of feeding and clothing its citizens) from biodiversity, but this value would never have been reported. This could be a problem for some third world nations, where a good fraction of the population lives directly off the natural resources of the country.

The second type of direct economic value, *productive use value,* is perhaps easier for us to understand because we deal more directly with it in our daily lives. Productive use value is the monetary value of things bought or sold. In terms of biodiversity, this would represent the amount of money you could get from taking some natural product directly out of the wild (e.g., redwood tree, wild parrot, or yellowfin tuna) and selling it on the open market. This is a very common reason people give for valuing biodiversity: namely, that we can convert products of the natural world into money. It has been estimated that, for the United States alone, the productive use value of natural resources is somewhere around $87 billion per year. Clearly, with numbers like this, it is easy to see why some people value biodiversity in this way, because there can be a huge economic incentive to maintain selected animals and plants, as well as the ecosystems that keep their populations going.

INDIRECT ECONOMIC VALUE

So far, we have been thinking about the immediate economic advantage to having species-rich ecosystems, but there are other economic issues as well. These are sometimes called *indirect economic values,* or future option values, and they deal with how much a particular resource will be worth to humans in the future. For example, we can look at this in terms of our agricultural or genetic resources. The majority of agricultural staples in the United States (corn, wheat, beans, potatoes, coffee, cattle, etc.) are the result of many generations of intense artificial selection. Our modern versions of crops and livestock have been bred to contain desired suites of traits and, as a result, are often grown as huge monocultural clones, meaning that all the individuals are essentially genetically identical. This can become a serious issue when disease breaks out or when pest infestations sweep through domesticated populations, all of which have the same lack of immunity. Suddenly, the ancestral or wild populations become of immense value because the entire genetic legacy of the

organism, including genetic variations for disease and pest resistance, is contained in the varieties found in nature (or in "primitive" cultivars). These wild-type traits may be selectively bred back into our crops, making them more genetically fit. If we lose these wild congeners and the genetic diversity they contain, then we lose the ability to bolster our crops' fitness, which jeopardizes our food supply. The value of biodiversity for the future of our agricultural or genetic resources is hard to estimate, but it is presumably billions of dollars per year.

Similarly, we can look at medicinal drugs and pharmaceuticals because over 80 percent of our medicines are derived from natural plant and animal compounds or from analogs based on them. Common drugs such as morphine, ephedrine, digitalis, pilocarpine, and quinine are derived from plants. It's been estimated that over 1,200 chemical compounds extracted from rainforest plants are currently being tested as potential cancer treatments. In very real ways, diverse natural communities such as tropical rainforests act as gigantic chemical research facilities, where individual species of plants have been developing and testing chemical compounds for millions of years. Some of the chemicals are used to protect the organism from environmental extremes, whereas others are used as weapons of defense against other organisms. We have only to visit the forests and search out the chemicals and drugs that are useful to us. We can value the rich biodiversity of such places because they provide us easy access to a wide array of bioactive compounds, and the continued presence of this biodiversity has been estimated to be worth at least $40 billion per year.

These values are increasingly being recognized as important. Brazil, for example, has laws in place that have a goal of no net loss of rainforest habitat, and other tropical countries are increasingly expecting royalties from any discoveries of valuable compounds from rainforest organisms. The transitory value of such discoveries is reflected in the value of the yew tree in the western United States. The yew is normally regarded as a minor tree in the forests, and as being of little value. Yet, when a compound (taxol) was extracted from the bark of the yew that proved to be effective in the treatment of breast cancer, suddenly people were searching all parts of the forests for the trees. Once taxol had been synthesized, however, the yew tree went back to being a low-value tree. Thus, a problem with indirect economic values is the difficulty of translating the economic benefits to people who have to live from day to day in or near the resource areas. A peasant illegally clearing an acre of rainforest to feed his family has little concern for the low probability that one of the plants he is burning might contain a cure for pancreatic cancer. Nevertheless, new screening methods for discovering potential drugs in wild organisms suggest that there is almost unlimited potential in nature, which will undoubtedly lead to exciting new treatments and medicines.

ECOSYSTEM SERVICES

A third set of reasons to value biodiversity is concerned with the *ecosystem services* provided by intact, functioning, and biodiverse ecosystems. Despite the fact that we can describe detailed food webs and measure the amount of energy that flows through an ecosystem, we ecologists don't fully understand how communities and ecosystems work. The problem has been described using the metaphor of a watch. At one time (prior to digital watches), watches were constructed using an elaborate set of gears and springs (Figure 8.2). The normal person could, using some simple tools, take apart all the pieces and lay them out on a table, but he or she would have little idea what each one of the tiny parts did, let alone be able to put them all back together in proper working order. It would be foolish to lose or discard any of the parts and expect the watch ever to tell time properly again. Ecosystems are like an amazingly complicated mechanical watch. We can identify and describe some portion of the parts, but we don't really understand how they all fit together. Throwing out parts or causing the extinction of species may be a foolish idea, because although we

FIGURE 8.2. Ecosystem services have been likened to a mechanical pocket watch. We may be able to disassemble the watch and examine some of the separate parts that make up the whole, but most of us would have a very hard time knowing how each part functions or how the intricate parts all fit together to work properly. If we want the watch to perform its function (i.e., to keep good time), then it would be foolish to discard any parts of the watch, no matter how tiny and seemingly insignificant, because we don't really understand which parts are vital. If the ecosystem is the functional watch, and the mechanical parts are individual species, then we can see why it is important to value every species for the role it plays in a functioning ecosystem. (Photo by M. P. Marchetti.)

may be able to lose some species with little or no consequences, ultimately we do not know which species will be vital to the functioning of the whole system.

Functional watches accurately record the passage of time for us; our intact functional ecosystems provide us with a host of services that we take for granted or don't even fully understand. Some of these ecosystem services are extremely important, such as cleaning our air (i.e., removing carbon dioxide and providing oxygen) and water. Nature is so good at cleaning water that many of our sewage treatment plants are turning to the use of natural wetlands and so-called green filters to clean and disinfect water before it goes back into rivers, where it is often diverted again for drinking water. If we disrupt the large natural systems on the planet, we are likely to feel the consequences somewhere down the line.

Provision of drinking water is, in fact, one of most valuable and most unrecognized ecosystem services we depend upon. Ultimately, all drinking water starts as precipitation and runoff from the ground, and we benefit from ecosystems that store it for us, either directly in lakes, meadows, and glaciers or indirectly in groundwater. Water that runs rapidly off denuded landscapes is unlikely to percolate into the ground and often must be captured for human use with expensive dams and reservoirs. Water quality also suffers in the process because water processed through natural systems is likely to contain fewer pollutants and other harmful substances. As international water issue expert Sandra Postel points out, our failure to protect water-producing ecosystems is increasing water scarcity and reducing food security, social stability, and chances of peace among nations. Protecting ecosystems and the vital services they provide may prevent future wars over water!

Intact natural ecosystems also have a large impact on major climatic features, including the weather. We saw in Chapter 4 on climate that the physical presence of a rainforest (i.e., respiration by trees and plants) helps provide the moisture that becomes the rain, which in turn makes the forest grow. Fewer trees equal less rain. Intact ecosystems also buffer us from extreme events such as droughts and floods. When we completely remove the plant community from steep hillsides, we often get catastrophic floods and mudslides when it rains. Why? The plants stabilize the soil, physically holding it in place and allowing it to act as a natural sponge that absorbs some of the rainwater. The consequences of not maintaining the forests and soil can be severe. Since the late 1980s, a series of major floods in Bangladesh have killed thousands of people, partly as a result of the intense logging in the upper watershed of the Ganges River in distant Nepal. Removal of the trees has increased the rapidity of water running down the hillsides, so the rivers have higher flows during floods (and lower flows later in the year).

Intact ecosystems also act as environmental monitors that can alert us when things are going wrong. You may have heard the phrase "a canary in a coal mine," which applies here. The idea comes from mine workers, who would take a canary into the mine with them in order to alert the miners of poor air quality, because canaries have low tolerance for "bad" air. If the canary died, then the miners knew it was time to get out before they suffocated. People have suggested that the presence or absence of certain species can act as warning signs of environmental damage. For example, amphibians, which absorb a large amount of their oxygen through their skin, are sensitive indicators of air quality changes. When amphibians decline or disappear, as has been happening around the globe, it is possible that atmospheric disturbances such as excess UV radiation or airborne pesticides may be the cause. In California, frog extirpations in remote mountain areas have been linked to airborne pesticides from large-scale agriculture, which suggests that people living or recreating in the areas are also inhaling these pesticides, with consequences for their health (and health costs). Intact ecosystems and their associated biodiversity can be valued for the environmental services they provide to humans, and estimates have been made that our ecosystem services are worth trillions of dollars annually. These services are truly something for all of us to value and protect.

INTRINSIC VALUE

A fourth set of reasons to value biodiversity is a radical departure from the sets of values we already have discussed. This value system for biodiversity shifts the focus from what biodiversity can do for us to recognizing the inherent value in biodiversity for itself. This is often referred to as *intrinsic value* and suggests that individual organisms have a right to exist on their own, without any relation to what you or I or any other person thinks about them. Recognizing the intrinsic worth of nature moves the rights and values of living things away from human desires (an anthropocentric view) toward inherent rights that organisms possess by the very fact that they are alive (a biocentric

FIGURE 8.3. An intrinsic value approach to the natural world suggests that all living creatures, irrespective of their size, utility for humans, or characteristics, have inherent value in and of themselves and that human beings do not have the right to cause their extinction. (Photo by C. Wayne Summers, Alexandria, Virginia.)

view). This view also moves humans from being the only important species and the obvious center of focus to being one species among many in a planet-wide community, all having value (Figure 8.3). With this set of values, humankind is placed firmly in the natural world, not as something unique or somehow separate from nature. We as humans are viewed as members of ecosystems that build and fuel the cycles of life on the planet.

It is also assumed under this set of views that humans do not have the right to cause the extinction of other species, because these species possess an inherent right to exist. Therefore, we are never justified in ending their existence. Interestingly enough, this concept is the basis for the federal Endangered Species Act, which says that it is the policy of the United States not to let any species go extinct. Some people have suggested a radical extension of this argument to include the idea that communities, ecosystems, and even the whole planet also possess inherent rights and value. This is sometimes seen as part of what is known as the Gaia hypothesis, popularized in the 1960s and 1970s by James Lovelock. Interestingly, in 2008, the country of Ecuador adopted a new version of its constitution that expressly recognizes that nature and natural processes have a right to exist. This is the first time any country has included language that expressly declares the intrinsic value of the

natural world; in time, it is hoped that more countries will follow Ecuador's lead.

One of the most eloquent people to first speak and write about the intrinsic value of nature was Aldo Leopold, who many consider a philosophical father for the discipline of conservation biology. In his inspired book, *A Sand County Almanac*, Leopold describes the development of what he calls a "land ethic," a philosophical stance in which the land and its inhabitants are considered to have inherent value. He writes, "a land ethic changes the role of *Homo sapiens* from conqueror of the land to plain member and citizen of it. It implies respect for the fellow members and respect for the whole community." Many people believe that the inherent value possessed by other living creatures is more than enough reason to protect biodiversity.

SPIRITUAL VALUES

A final set of reasons to value biodiversity arises from the human tendency to have religious and spiritual beliefs. All of the major religious traditions of the world (e.g., Hinduism, Islam, Buddhism, Judaism, Christianity) contain lines of thought and discourse whereby humans are given a role in protecting or respecting nature or other forms of life. This is also often a large part of many of the less well known, so-called traditional, spiritual belief systems (e.g., many Native American belief systems) (Figure 8.4). It is often suggested in these belief systems that humans have a responsibility, sometimes divinely given, to take care of the planet. Alternatively, some religious practices teach that humans have a responsibility to future generations to leave the Earth and its inhabitants in good condition. In the Judeo-Christian tradition, for example, there have been mainline theologians, such as Francis of Assisi, who called for human beings to be good stewards of the Earth and to protect and cherish the animals and plants given to them by a divine being. Good stewards do not waste and destroy the precious gifts entrusted into their care; instead, they nurture and defend them. Increasingly, Christian churches in the United

FIGURE 8.4. Temple I (Temple of the Giant Jaguar) in the Mayan ruins of Tikal, Guatemala. Most religious traditions place some amount of spiritual value on nature and the natural world. (Photo by C. Wayne Summers, Alexandria Virginia.)

States are suggesting that their members take more of an active role in protecting other forms of life by doing things such as planting trees or changing their buying habits. Spiritual values can be a powerful force in motivating people to change their behavior and to place some value on the biodiversity of nature.

As Gary Gardner, a senior researcher at the Worldwatch Institute, points out, religions "shape people's world views, wield moral authority, have the ear of multitudes of adherents, often possess strong financial and institutional assets, and are strong generators of social capital, an asset in community building." Thus, religious organizations have powerful potential to lead humanity on the pathway to an environmentally sustainable planet. Many are already starting to doing so, guided by spiritual values. For example, efforts to clean up the Ganges River in India are being led by the Clean Ganga Campaign, which has strong roots in traditional Hinduism and is headed by V. K. Mishra, who is both a scientist and a priest.

ECONOMICS

As we have seen above, monetary or economic values are often among the most powerful motivating factors for many people in the western world. Therefore, a basic knowledge of economics can help us form linkages between disciplines (e.g., ecology, philosophy, sociology, conservation) and provide us with a better understanding of the hows and whys behind economic transactions and their effects on biodiversity.

Economics is defined as the study of production, distribution, and consumption of goods and services as related to human beings. Historically, economists have tried to answer questions such as the following:

What types of goods should be produced?

How much should be produced?

In what manner should the goods be produced?

How should the goods be allocated?

One of the major schools of thought within economics is classical economic theory. This theory describes the market as a force, which is considered to be the sum total of everyone's individual needs and wants. When people act in their own self-interest (i.e., when they do the things necessary to meet their needs), then we have what is called a free market, which is reflective and responsive to people's demands. Under this theory, a stable and prosperous economy is reached when market forces are free to guide and direct the economy in meeting people's needs.

Classical economic theory can also be seen as a framework in which to interpret the way the world works (also known as a worldview) and can be represented by the top image of Figure 8.5. In this representation, there are three general parts: the extractive sector, which gets the raw materials; the production sector, which translates the raw materials into goods or services; and the consumptive sector, which consumes the products.

There are a number of important things to notice about classical economic theory as a framework. First, it is a very limited view of the world and is completely human centered. For example, there is no natural environment (i.e., no plants or animals) in this schema; the economy is all that there is. This worldview also

Classical Economic Worldview

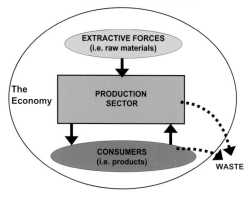

Neoclassical Economic Worldview

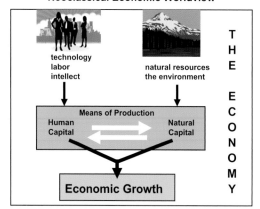

FIGURE 8.5. In the classical economic worldview on the top, the economy is all there is, and it is represented in the figure by the black circle. Note that the environment is not considered to be part of the economy and that waste from economic activity is transported out of the system. In the neoclassical economic worldview, the environment is included as natural capital in the economy, but it is equated with human capital, suggesting that human ingenuity can solve any environmental shortcoming. (Modified after figures from Van Dyke, F. 2003. Conservation biology: foundations, concepts, applications. McGraw-Hill. New York.)

postulates a closed system, where materials used in the economy come from somewhere outside the system, and waste from production and consumption is exported again to somewhere outside the system. The only things of real value in this scheme are ones that increase production, and which thereby generate more money and ultimately satisfy more people's wants and desires. Most importantly

in a classical economic framework, there are no real limits to growth. The extractive sector can always get more raw materials (i.e., they are limitless), the productive sector will always find buyers for its products, and wastes will always be removed from the system and cause no harm.

As the field of economics progressed, an additional body of thought called neoclassical economic theory developed, partly as a response to some of the shortcomings of classical economic theory. Neoclassical theory can be represented by the bottom image of Figure 8.5. Notice that this worldview is a bit of an improvement from classical theory. The new theory recognizes two forces fueling production: the environment, as the place where resources originate, and the human work force, in terms of technology, labor, and human intellect. Also, the productive sector is now thought of in terms of both natural capital (i.e., natural resources) and human capital (i.e., human resources).

There are still some large problems with this view of the world, however. First, although the natural environment is recognized, it is only present as a subset of the economy, and its only purpose is to feed the means of production. In addition, natural capital is directly equated with human capital, which suggests that the two are interchangeable. This means that when resources (i.e., natural capital) become scarce, the economy will be able to substitute labor or technology (i.e., human capital). In other words, shortages of natural resources will be solved by both technology and human ingenuity or by an influx of human labor. Again, in this system of thinking, there is no limit to growth, human capital is interchangeable with natural capital, and technology will always solve the world's problems.

CRITICISMS OF ECONOMIC WORLDVIEWS

Clearly, the above descriptions are gross oversimplifications, but they help us to understand some of the problems for conservation that are inherent in dominant economic worldviews. So,

TABLE 8.1
External Costs of Fossil Fuels

EXTERNALITY (HIDDEN COST)	SYSTEM AFFECTED
Acid rain and ozone	Forests and forest ecology globally; lakes and streams
CO_2 and methane (greenhouse gases)	Global climate, ocean productivity
Air quality, asthma, cancer rates, etc.	Human health and associated costs

in addition to the difficulties recognized above, there are other fundamental problems with using economic theory as a way to organize and direct human society. First and foremost, the standard economic theories used in western nations suggest that the economy must continue to grow and that growth is unlimited. Unfortunately for us, this goes against everything we know about how the world actually works. Nothing in nature increases forever or grows unchecked, because there are always limits. The limits can be physical limits, as in the finite amount of natural gas on the planet, or they can be ecological limits, as in the fact that populations do not grow exponentially forever (think about carrying capacity). Basing all of our society and its future on a belief system that posits infinite growth is inherently irrational. Nothing in nature grows forever, and neither can the economy (or the human population).

Another criticism of economic theory, similar to the one above, comes from the fact that natural resources are never infinite. We know that natural resources, such as species and their entire genetic legacy, can be lost forever from the planet. We also know that one species is not equivalent to another species and that we cannot change this by use of our technology or ingenuity. We therefore can not substitute the introduced European starling for the extinct native passenger pigeon; nature does not work that way. When a species is extinct, it is gone forever.

Also, many of our renewable natural resources are not renewable on a time scale useful for human beings, as is the case with fossil fuels. Fossil fuels are made from natural products, but it takes millions of years to produce a gallon of oil. So in a very real sense, the growth of our modern civilization based on an oil economy is literally burning up millions of years of accumulated energy over a tiny amount of time. It's also possible that feedback loops that provide our natural ecosystem services with resilience may fail long before a finite natural resource such as tropical rainforests is completely gone.

A third and particularly insidious problem with modern economic thought is the fact that our economic systems mask the true cost of production through what are known as externalities. *An externality is a hidden cost to the economy that is shared by everyone.* An example of this is in Table 8.1, which lists some of the external costs associated with the use of fossil fuels as our main source of energy. Our economic system generates a price for a gallon of gas based partially on the amount of money it costs to extract it, refine it, and ship it. The cost of gas is also reflective of how much people want or need it (e.g., if no one drove cars, the price for gas would initially drop because of a lack of demand). We, the consumers, pay for all these upfront costs when we fuel our cars. We are basically paying for someone (i.e., oil companies) to remove the resource from the ground, clean it up, and get it to our neighborhood gas station. The problem is that there are hidden costs (externalities) associated with every gallon of gas we buy that are not reflected in the price we pay at the pump.

The hidden costs to society can be seen in issues such as human health. Burning fossil fuels increases the amount of air pollution and smog, which has a direct impact on the rates of asthma, respiratory disease, and cancer. We, as members of this society, all have to pay the price of this hidden cost in terms of our health insurance and health care (and we have to breathe the polluted air). For example, if you live in a city with perpetual bad air quality, you will have to pay your health insurance company more each month to cover the increased payouts the company is going to have to make, because you are more likely to get sick. And when people in the same location don't have private health insurance, we as citizens all pay for the costs of their public health insurance through our taxes (or for the cost of emergency treatment, in the case of those who are uninsured, through our medical payments). The burning of fossil fuels is a cost to all of us, but we generally don't ever examine in any detail what this means to us individually.

Other externalities associated with fossil fuel use include damage to the Earth's climate (i.e., global warming) and the associated environmental, biodiversity, and ecosystem service losses. If we were to calculate the total cost of a gallon of gas, and we included in the price the amount of money needed to fix or address these hidden problems, we would likely be paying double or even triple for each gallon. But our economic system does not see issues such as wastes (e.g., carbon dioxide) as part of the economy and therefore excludes these problems when setting prices. Externalities are more than just abstractions for economists to fret over; they represent the major environmental problems we face as a species. Our economic systems are, in effect, causing current environmental disaster and biodiversity collapse because we are not paying for the damage we do to nature. This behavior stems in part from what the late Dr. Garrett Hardin called "the tragedy of the commons."

TRAGEDY OF THE COMMONS

To get a feel for the tragedy of the commons, let's walk through a short thought experiment. Suppose you are one of 10 families living on a small island in the South Pacific. Your community decided decades ago to pool its resources, in terms of grazing land for cattle, so that there is one common pasture area where everyone grazes his or her cattle. At the time this group policy was enacted, it looked like there were enough resources in the pasture for each family to raise five cows per year, so every family was limited to grazing five cows and no more. You have all been following this guideline for a good number of years, and everyone is doing okay until two things happen in the same year to your family: you get a bad case of avian influenza and cannot work the farm for a few months, and your wife unexpectedly gives birth to triplets. Now, unfortunately, the bills are starting to pile up, and it looks like you may not have enough food for the whole year. Thus, being a smart farmer, you sneak an extra cow onto the pasture, thinking, What is one cow among 50? No one will notice, and you will have the extra resources you need to feed and clothe your family for a year. It sounds logical, so you do it.

Luckily for you, even with the extra difficulties, your worries for the year were unfounded, and you had enough resources to feed your family. The extra cow was not needed after all, and it essentially became profit for you and your family. With the extra money the cow brought you, you buy a new plow, a new stove for the house, and a Nintendo system for the kids. The next year, you decide to sneak an extra cow onto the pasture again, because, really, what can it hurt? But three of your neighbors noticed your behavior, saw the new things you bought, and decided to follow suit and also slip an extra cow onto the common field. This goes on for a while, and eventually the majority of the families notice and also begin to regularly add cows to the field, some years one extra, some years two or three extra, until, on average, there are about 100 cows a year on the pasture that used to sustain 50. The community is becoming prosperous with all this added wealth, and everyone is happy, until the rains do not come for two years in a row. Now, the field that once easily supported the community's population of cows through droughts in the past cannot sustain twice the added pressure, and the whole system collapses (i.e., you exceeded the pasture's carrying capacity, or k). The majority of the cows die of starvation over a period of weeks, and many of the original 10 families are forced to leave the island for good.

What happened here? What was the problem, and what caused it? Everyone was doing the appropriate thing in looking out for himself or herself and his or her family. Yet eventually there was an environmental disaster. How is it that rational self-interest and wise use of a resource can lead to such big problems? The scenario described above has variously been labeled "the tragedy of the commons" or "the problem of common access resources." It has been suggested that situations analogous to this are at the root of the vast majority of the environmental problems we face today.

At its basic level, the tragedy of the commons is a conflict over resources between individual self-interest and the common good. The metaphor of the common pasture illustrates how unrestricted access to a limited resource eventually dooms the resource through overexploitation. This occurs because the benefits of using the resource accrue to individuals, each of whom is trying to rationally maximize his or her gain, whereas the costs of the overexploitation are distributed among all the people who have access to the resource (which may be a larger group than those who are exploiting it). You can imagine that this kind of economic problem would be difficult to solve because there is an incentive for the individual to continue to overexploit the resource in every step of the process. The selfish or self-serving individual always comes out ahead because the cost he or she inflicts is spread over the entire population, whereas the benefit goes only to the selfish user.

The idea of the tragedy of the commons has been applied to many environmental problems where there is a common resource such as air pollution, water pollution, overextraction of fresh water from aquifers, soil contamination, grazing on public lands, mining on public lands, logging on public lands, overfishing of the oceans, and burning of fossil fuels. Ultimately, the tragedy will affect the global commons through human overpopulation of the planet, given our finite set of global resources.

Popularization of the ideas of common access resource problems began with a 1968 article by Dr. Garret Hardin in the journal *Science*. In the article, Hardin described the problem and posed several possible solutions. One of the easiest solutions to this set of problems involves a "big-stick model" of regulation. If the common resource is managed or policed or governed by a body that has strong enforcement powers (i.e., a "big stick"), then the selfish parties can be discouraged from overuse. This type of enforcement can be financial, involving levying a fine, tax, or usage or permit fee, as long as the amount is large enough to actually discourage overuse. Additionally, the governing body also needs to have enough political or economic clout to enforce the fines or other punishments, which is not always the case.

A second potential solution to the tragedy of the commons lies with education. If the population as a whole is well educated about the problems of overuse, then potential overexploiters may think twice about a particular selfish action. People have variously called the tragedy of the commons a problem with human greed, cheating, or selfishness, but given the proper environment and education, many people will choose to be somewhat altruistic and to work toward a common good. Selfishness does not always need to win, and education can help tremendously in that battle.

A third type of solution relies on education but does so in the context of small communities. When the group or community size is small and everyone knows everyone else, it is harder to go against the group's wishes and easier to do the right thing. Small groups are much better at self-regulating and can notice and punish cheaters faster than can large, anonymous groups. In small communities, the good of the individual is often tied more directly to the good of the whole collective, so a small group can agree to cooperate in the name of mutual benefit. The idea of living and working in small communities may be a partial solution to many of the environmental problems people face today.

ECOLOGICAL ECONOMICS

It would seem from the above discussions that western society is perhaps doomed to environmental failure due to blind adherence to, and overwhelming success of, our economic system of values. Thankfully, in the last few decades, many economists have been reexamining the basic tenants of the western market economic system and have proposed new economic models based on environmental or ecological logic. Ecological economics presents a pluralistic approach to the solution of economic problems, with a focus on long-term environmental sustainability.

One of the objectives of ecological economics is to remake economic theory based on the constraints of physical reality, particularly the limits to the environment and the laws of physics. If we use our economic system to plan for ecological and sustainable development, then we can improve our well being and limit the size of the human footprint on the planet. Some ecological economists assert that the field of economics should be seen as a subdiscipline of ecology, because ecology deals with the interactions among living beings and nonliving matter on the planet, and the human economy is, by definition, contained within the planet.

A true environmental economy would appear in stark contrast to both the classical and neoclassical economic worldviews, in that both of these theories assert that infinite economic growth is both desirable and possible. In contrast, ecological economics depends on knowledge of ecological systems and recognizes the important contribution that natural capital, such as renewable runs of salmon, functioning wetlands, and old-growth forests, makes to economic systems. Our extraordinarily intricate and diverse ecosystems can be seen as natural capital in its most basic form. Thus, functioning and intact natural ecosystems act to build our natural capital and provide services and products directly to human societies. In addition, ecological economics asserts that not only is human capital not equal to natural capital but also that human capital is in fact completely dependent on natural capital. That is, without our ecological support systems, there would be no humans. A famous study by economist Robert Costanza and others in 1997 estimated that the global value of just 17 ecological services was twice the value of the entire "real" global economy. An additional study in 2002 calculated the economic value of wild lands to humans at over $250 billion *per year,* clearly suggesting that we rely very heavily on our natural environment.

Ecological economics and economic systems that include ideas of sustainability are gaining increasing attention around the world. Some interesting ideas are coming out of these new schools of thought and include an attempt to gauge people's interest in solving environmental issues with a measure called "willingness to pay." This attempts to clarify how important environmental and ecological services are to people by asking them how much they would be willing to pay to save or preserve some ecological feature. For example, surveys have been sent out asking the general public how much it would be willing to give in order to ensure the existence of intact rainforests for the next generation or to secure the presence of charismatic species such as pandas. Another idea is to use what is called a "travel cost method," in which an assessment is made as to how far a person is willing to travel to visit a place such as a coral reef or rainforest. Results from these types of evaluations suggest that environmental protection is becoming increasingly important to the general public.

Another idea coming out of these schools of thought is an idea termed "polluter pays," whereby polluters, not society, pay in advance for the cost of cleaning up their pollution. This means that society would receive the true cost from the user of the environment and that externalities would no longer be hidden and passed on to the general public. For example, this would translate to more expensive gas and energy for all of us, because we would pay up front for the true cost of using fossil fuels. But this would also provide a strong economic incentive to not drive needlessly or to otherwise waste energy. People who were more sustainable in their use would strongly benefit, whereas those who overindulged and were wasteful would pay the true cost of their energy use.

Some respected economists such as Herman Daly of the University of Maryland advocate even more radical ideas, such as a large-scale redistribution of resources and extreme limits to population growth. Daly has strongly questioned the value of acquiring a large amount of possessions and wealth. In his arguments, he notes that psychological studies have shown that objects do not equate with happiness and that much of what is important in human well being is not analyzable from a strictly economic standpoint. In the long run, it appears that intangible elements such as "quality of life" are likely to be more important to human happiness than are mere objects, possessions, and toys. Future economic systems may be based on trying to balance intangible quality-of-life issues with sustainable development and growth. We will know ecological economics has finally become acceptable to the world at large when Daly and Robert Costanza are awarded a Nobel Prize for economics (an unlikely possibility).

"When iPods are valued over whale pods, the economic system will deliver ever more species of iPods and wipe out yet another species of whale" (Bayon 2008).

FURTHER READING

Note that a majority of the references below come from the Worldwatch Institute. This institute's publications on a wide array of subjects related to the environment and economy are accessible and inexpensive.

Balmford, A., et al. 2002. Economic reasons for conserving wild nature. Science 297(5583):950–953.

Bayon, R. 2008. Banking on biodiversity. State of the world 2008: innovations for a sustainable economy (pages 123–137). Worldwatch Institute. Washington, DC.

Costanza, R., et al. 1997. The value of the world's ecosystem services and natural capital. Nature 387:253–260.

Daily, G. C., ed. 1997. Nature's services: societal dependence on natural ecosystems. Island Press. Washington, DC.

Daly, H., and J. Farley. 2004. Ecological economics: principles and applications. Island Press. Washington, DC.

Gardner, G. 2002. Invoking the spirit: religion and spirituality in the quest for a sustainable world. Worldwatch Paper 164. Worldwatch Institute. Washington, DC.

Gardner, G., and T. Prugh. 2008. Seeding the sustainable economy. State of the world 2008: innovations for a sustainable economy (pages 3–17). Worldwatch Institute. Washington, DC.

Hardin, G. 1968. The tragedy of the commons. Science 162:1243–1248.

Postel, S. 1996. Dividing the waters: food security, ecosystem health, and the new politics of scarcity. Worldwatch Paper 132. Worldwatch Institute. Washington, DC.

9

Conservation Science

The appeal of *conservation science* is that it is a truly integrative discipline focused on understanding how humans are changing the world and on finding practical solutions to protecting biodiversity. What we call conservation science is usually called *conservation biology,* but we are using a broader name to reflect the many disciplines that contribute to it. Its inspirational roots are in the environmental movement; its theory derives from the sciences, especially ecology and evolutionary biology; its concerns for the needs of people come from the social sciences and humanities; and its practical orientation has roots in wildlife management, forestry, and similar applied disciplines. Increasingly, the physical sciences (i.e., physics, chemistry, etc.) are also contributing to the field by developing better ways of understanding many of the physical processes that contribute to global climate change (e.g., currents in the air and oceans). The discipline of conservation science increasingly draws support from the public concern over the effects of environmental change on human health and well being. Thus, practitioners recognize that the philosophical and spiritual roots of conservation science lie in the world's religious and literary traditions, and increasingly practitioners work in non-biological fields such as politics, law, and economics. Arguably, the goal of conservation science is to make the planet sustainable, creating a situation where humans and the Earth's rich diversity of life share the future.

Although conservation science is ultimately a global science, from a practical perspective, conservation takes place at three basic levels: the population or species level, the habitat or community level, and the level of entire landscapes and ecosystems. In the following sections, we will address the tools and approaches used by conservation scientists at each of these levels.

SPECIES-LEVEL CONSERVATION

Biologists have long realized that small populations are more susceptible to extinction than large populations. This is hardly surprising, but it turns out to be an extremely important fact to keep in mind for conservation at the level of individual species. Small populations are likely

FIGURE 9.1. The idea of a genetic bottleneck is illustrated using a jar of blue- and green-colored beans. The bottle could represent a population of an animal species with two color morphs, dark and light, controlled by a single gene with two alleles (dark and light). See text for complete explanation. (Figure by M. P. Marchetti.)

Original population has equal #'s of the two color alleles

An environmental event reduces the population & only a few survive

By chance the survivors are not made up of equal #'s of the two alleles

So the next generation has more blue than green alleles

to go extinct for one of two general reasons: genetic issues and random events.

GENETIC ISSUES

As we have seen, natural variation is a vital component of evolutionary fitness, because it is the differences among individuals that allow natural selection to work. Small populations lose much of their natural variation just by being small, because genetic diversity is usually lost when a large population becomes smaller. In a shrinking population, genetic variation can become so small that it can reduce a species' adaptability, even if the population grows large again. This phenomenon is known as a *genetic bottleneck*. Figure 9.1 shows how this could work using beans in a jar. In this cartoon system, individuals are the beans, and genetic diversity is represented by the two different colors (think two different alleles). Our starting population has an equal frequency of both colors (50 percent are dark and 50 percent are light). Let's pretend that our population has a very bad year and that all but 10 individuals (i.e., beans) are killed in a freak tornado (represented in the figure as pouring 10 beans from the bottle). By chance, eight of the ten that survive the storm are dark colored. When the population numbers rebound following the tornado, notice that the proportion of dark to light is skewed heavily toward the dark beans, and therefore that the overall genetic diversity of the population has decreased. In this system, we were only looking at one genetic

trait (bean color), but in living organisms, this same process can result in dramatic losses of genetic variation over hundreds of thousands of genes. In other words, small population size can negatively affect the genetic diversity of a population. With less genetic material, populations may lose their ability to adapt to changes in the environment, have reduced fertility, and become more susceptible to genetically related diseases and other problems.

Another genetic issue with small populations is the problem of inbreeding. If individuals in a small population only mate with each other, then the incidence of rare genetic diseases can increase dramatically within the whole population. This is called *inbreeding depression*. Inbreeding depression occurred in many of the royal families in Europe, particularly among the ruling dynasties of Spain and Portugal. To preserve the royal bloodlines individuals, were urged to marry cousins and close relatives. Over time, many royal families started developing rare genetic diseases as the result of inbreeding, such as hemophilia and mental deficiencies. Hemophilia is a disease where the clotting factors in the blood that allow cuts to stop bleeding don't function properly. As a result, even small cuts can be life threatening. The royal families presumably suffered reduced genetic fitness as a result of their close marriages.

This same thing can happen in plant and animal populations as well, with the classic example being inbreeding problems in African cheetah populations. The cheetah occurs naturally in the African savannah in low population numbers. As a result, cheetahs have been found to have extremely low genetic variability, indicating that all cheetahs today are most likely descended from a very small population that experienced a genetic bottleneck in the past. There is concern that the low genetic diversity of cheetahs may make them exceptionally vulnerable to epidemic diseases or may make it difficult for them to adapt to major climatic changes in the future.

RANDOM EVENTS

Unexpected or unpredictable changes in the environment can have dramatic effects on population numbers. We have seen a clear example of this in the chapter on natural selection, when we looked at the Galapagos finch population's decline following a drought. These types of environmental fluctuations (i.e., droughts, floods, hurricanes, fires, etc.) are a natural part of every ecosystem. But if a species has a low number of individuals already, then a natural disturbance can, in the blink of an eye, wipe it out entirely. Having a small population size puts a species at a much higher risk of extinction due to chance events. One of the major effects of human domination of the Earth is that the frequency of events that can wipe out small populations has greatly increased. As we have witnessed, numerous wildlife populations are threatened by human-caused "events," such as oil spills, toxic waste, and even nuclear fallout. When the number of individuals in a population is low, the chance of extinction due to an accident, including natural events such as a hurricane, is high. In short, low population numbers greatly increase the probability of extinction from stochastic events.

PVA AND MVP

If small population size is bad for long term survival, then the big question becomes, How small can a population become before its extinction is assured? In attempts to answer this, conservation biologists have turned to *population viability*

analysis (PVA). PVA is a statistical method used to assess the risk of extinction. In other words, PVA is an attempt to determine if a species has the ability to persist in a given environment. The PVA process looks at the range and availability of resources that a species requires and tries to determine if a particular population of that species is likely to be viable in its environment. This is a very difficult and time-consuming process, requiring extensive fieldwork, data collection, and computer simulation models. For example, a PVA often needs to determine a suite of population-level traits, including birth and death rates, the sex ratio for the population, population size, and population growth rate, as well as causes of the species decline, amount of available habitat, and many other parameters. When all the data are collected for these parameters, then a population growth model is constructed (similar to the models we saw in Chapter 5), and the population size for the species can then be projected at various times in the future. Each PVA is developed individually for a target species or population, and as a result, each PVA is unique. When conducting a PVA, the goal is to determine if the at-risk population can become self-sustaining over the long term or if extinction is likely or imminent.

One of the measures used in a PVA is a determination of the *minimum viable population size,* or MVP. The MVP has been defined as the smallest population size that has a 99 percent chance of not going extinct for 1,000 years. In other words, it is the lowest number of individuals that can maintain a population for the foreseeable future. Determining a minimum viable population is also not an easy task and often involves the use of complex models based on the logistic growth model we have already seen. For example, an MVP analysis, based on population models, was used to assess grizzly bear *(Ursus arctos horribilis)* survival in Yellowstone National Park. It was found that the grizzly bear's survival was most affected by factors such as death rate, sex ratio, age at first reproduction, and territory size. Specifically, the analysis showed that populations smaller than 30–70 bears living in an area less than 2,500–7,400 square kilometers had less than a 5 percent chance of surviving 100 years! In other words, to maintain a viable population of grizzly bears, a very large bear population is required, coupled with the ability for the bears to range over an enormous geographic area, much bigger than Yellowstone National Park.

Ideally, when determining an MVP, a conservation scientist will weigh both the requirements of an individual species and the external factors that impact its ability to persist in the future. Factors that test the abilities of a population to persist include natural factors, such as variation in demographic, environmental, and genetic forces, as well as natural catastrophes. The more frequently a population is exposed to these events, the more likely it is to become extinct.

Human factors are often an added strain on species already stressed and leaning toward extinction. For example, the combination of natural and human factors has caused coho salmon (*Oncorhynchus kisutch*) populations, which once supported important fisheries, to reach dangerously low levels in the coastal streams of California and Oregon. Natural factors contributing to this include long-term droughts and major floods, which have reduced habitat for juveniles, and changes in the open ocean, which have reduced the survival of adults. Contributing human factors include activities such as logging, urbanization, and road-building, which degrade the salmon's freshwater habitats, as well as fishing pressure in the ocean, which reduces adult populations. California coho were listed as a threatened species by the National Marine Fisheries Service in 1997, closing the fishery. This listing was a recognition that many populations of coho were already extinct and that others were at or below the minimum viable population level. Not surprisingly, the remaining coho populations have continued to decline, and it is likely that only heroic efforts involving watershed restoration and captive rearing will keep them from extinction.

Because the factors that affect minimum population size are often intertwined and inherently difficult to quantify, assessing the relative importance of each is sometimes close to guesswork. For example, the heath hen *(Tympanuchus cupido)*, a kind of grouse, was once a common bird that ranged from New England to Virginia. Its total population was reduced to about 100 individuals on the island of Martha's Vineyard by 1900, as the result of human-caused habitat changes and hunting. A portion of the island was set aside as a refuge, and under close management, the bird's population increased to 800 in just 16 years. However, within just a few years, a series of unfortunate events (a large fire, predation by an unusually high number of goshawks, and disease) took its toll. By 1920 the population was again under 100 individuals, and 12 years later, the last survivor of the population (which had a high percentage of sterile males due to inbreeding) died. Eight hundred individuals seemed like a viable population, but the threshold for the minimum viable population was probably already passed when the population decreased to 100 birds in 1900.

SPECIES-LEVEL EFFORTS

So far, all we have done is describe some of the constraints and tools used when working on individual populations or species, but how do conservation scientists actually protect species? We will now look at two different approaches for species protection efforts, ex-situ conservation and in-situ conservation.

EX-SITU CONSERVATION

When a species has reached numbers close to what biologists judge is the minimum viable population size, ex-situ conservation programs may be initiated. The word *situ* in Latin means place, and therefore *ex-situ* conservation measures are defined by taking the species out of its place. We can think about this as conservation work that happens in a zoo or aquarium or botanical garden. In practice, ex-situ conservation efforts increasingly involve initiating captive breeding programs for animals or captive propagation for plants. The breeding of wild animals in zoos is practiced for many purposes, but recently the focus has shifted toward breeding endangered species with the intent of reintroducing them into the wild.

Most conservationists abhor the thought of placing wild animals in zoos or game parks, but as a last resort, it is preferable to the total annihilation of a species. In the past, zoos were seen by many as either places that abused animals through neglect or poor living conditions, or as environments that had little redeeming value as conservation tools. Captive breeding programs were scarce and often unsuccessful. In the early 1970s, it was found that, of 162 rare or endangered mammal species in U.S. zoos, 73 had been bred, but only about 30 had met with enough success to provide any hope for their reintroduction into the wild.

More recently, though, zoos and game parks have been receiving more favorable attention for their successes with captive propagation as well as their role in educating the public. For example, the Arabian oryx *(Oryx leucoryx)*, is a small, almost pure white antelope with long, nearly straight horns, native to desert and steppe areas of the Arabian Peninsula. Killing an oryx, known for its endurance and strength, was once considered a sign of manhood to the indigenous people in the area. This practice did not threaten oryx populations when men hunted with spears or single-shot rifles, but the introduction of automobiles and automatic weapons led quickly to the Arabian oryx's extirpation in the wild. Fortunately, the Arabian oryx existed in zoos; there were 64 living in three U.S. zoos in 1979. Since this time, Arabian oryx have been successfully bred and reintroduced into the Middle East, and with Bedouin guards protecting these populations (the people who were partly responsible for their demise), they appear to be doing quite well. Other successful captive breeding and reintroduction programs include the return of European bison to Poland, the black buck to Asia, and (perhaps) the California condor to the western United States.

Too often, however, success is claimed only when the species has been bred successfully in captivity, which is really just the first step in the preservation process. For example, if the reintroduced oryx never learn their wild ancestor's behavior of migrating to seasonal waterholes, then has the species really been saved? It is hard to separate a species from its surroundings and habitat and call it a successful conservation program. Issues such as this are often raised by captive breeding program critics, who cite genetic, ecological, and behavioral concerns. For example, there is a herd of eland *(Taurotragus oryx)*, a large African ungulate, introduced into the Askania Nova Biosphere Reserve in the Ukraine, that suffers from a high level of disease due to inbreeding. Or consider recent attempts to breed the European white stork *(Ciconia ciconia)*. Most of the white storks raised in captivity and released into the wild do not make winter migrations to Africa as the wild birds do. To ensure their survival, captive-raised storks must be fed throughout the winter months. In some cases, captive-reared storks are even displacing some of the few remaining wild pairs. However, climate change may also be playing a positive role here by creating conditions that allow storks to overwinter in areas, such as Spain and Portugal, which were not suitable in the past.

These examples raise a deeper philosophical question: when a species is approaching extinction, how much change in its genetic or behavioral traits are we willing to tolerate in order to save it? Frequently, when captive breeding is "successful," the species can be maintained only in an artificial environment. Over time, ex-situ conservation efforts select for traits that are adaptive in captivity rather than in the wild, so they potentially fail to maintain the full behavioral repertoire of the original wild species.

The inevitable effects of captive breeding, combined with a feeling for the intrinsic right of animals to remain "free," has led some conservationists to object to ex-situ conservation efforts on any grounds, because they fear that the practice will remove the impetus to preserve natural habitats. Others point out that the decision to conserve a rare animal through ex-situ conservation efforts is a decision to sacrifice a significant number of the few remaining wild specimens to a program that may fail. Despite the successes of ex-situ conservation efforts, they can never really be a panacea for extinction. This leads us to consider another means of saving species, which is through in-situ conservation work, or the protection of species in the habitats in which they live.

IN-SITU CONSERVATION

Habitat destruction is the number one cause of extinction, so if we truly want to protect a species, then we need to protect its habitat through in-situ conservation efforts in protected areas such as reserves, parks, and wilderness areas. But what exactly is a reserve? Reserves come in all shapes and sizes and are established for various purposes. Some protected areas (sometimes called preserves) strictly limit human activity, as is the case in designated wilderness areas within the United States. In these protected habitats, most human activities beyond hiking and backpacking are prohibited. For example, no motors are allowed to be used in wilderness areas, even for conservation work such as trail clearing and maintenance. Other protected areas permit various levels of human use. For example, *partial reserves* may be fully protected only during the breeding season of a target species. Thus, some beaches are closed during the breeding season of the least tern (a small fish-eating bird) and various sea turtles, whereas for the rest of the year they are open for recreation. Another possibility is what is known as an *extractive reserve,* which allows for the extraction of some resources in a carefully managed way to ensure the protection of the entire reserve ecosystem. Francisco Filho, aka "Chico Mendes," a famous Brazilian environmentalist, brought this type of system into the limelight in his quest to protect Amazonian rainforests for the extraction of rubber from wild rubber trees.

It has become increasingly apparent that large expanses of land relatively free from human pressures, where we can protect wild species, are difficult to find and establish. In recent years, some conservation practitioners have decided that the establishment of extractive or multi-use reserves is an acceptable compromise that allows humans and wildlife to coexist. Whatever type of reserve is established, it is best if the reserve is part of a broader system of protected areas that are under active management. Conservation biologist Michael Rosenzweig argues that natural areas that support biodiversity have to be integrated into the human-dominated landscape because the protected areas are too few and small (even when all added together) to do an adequate job of protecting biodiversity. Thus, one of the big questions around reserve design is, Given limited financial resources, what is the best way to establish a protected area? We will examine this question in detail later in this chapter.

COMMUNITY-LEVEL CONSERVATION

An alternative to a strictly species-level conservation approach is to focus protection efforts on an entire habitat or community. This approach has some definite advantages. If you protect a community, then you are essentially protecting an entire group of individual species, which is clearly less time-consuming than if you had to work on each species one at a time. In addition, we know from our survey of ecological ideas (Chapters 5 and 6) that individual species do not live in a vacuum; rather, they are linked to hundreds of other species through food webs and other ecological relationships. Therefore, if we truly want to preserve a species, we also need to protect its habitat and community. So why not focus our energies at this level? We will explore what it means to do conservation work with communities by first examining the major complicating factors involved in this kind of enterprise.

COMMUNITY-LEVEL COMPLICATIONS

There are major challenges facing any scientist working at the community level of ecological organization. First, it only takes a casual look at the world to notice that natural variation

FIGURE 9.2. Natural habitat variability as seen in the foothills of northern California. (Photo by M. P. Marchetti.)

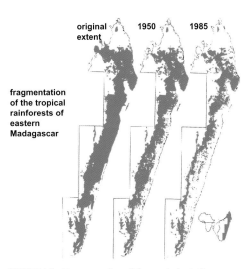

FIGURE 9.3. Fragmentation of the tropical rainforest of eastern Madagascar at three points in time. Note that what was once a continuous tract of rainforest has become disjointed and fragmented over its entire length. (Reproduced with permission from Sussman, R. W., G. M. Green, and L. K. Sussman. 1994. Satellite imagery, human ecology, anthropology, and deforestation in Madagascar. Human Ecology 22(3):333–354.)

is present throughout and across every community (Figure 9.2). Trees are not distributed evenly across a forest, fish are not equally spread out in a lake, and bears are not found in every cave in the mountains. All habitats are variable across space and are sometimes referred to as being patchy in nature. Given this constraint, the question becomes, How can we protect a community that is not the same everywhere it is found? A strategy that might work for the protection of a desert-floor stream community might not be appropriate for a stream higher in the mountains. Variability across communities increases the difficulty of conservation.

Secondly, when we studied communities in the ecology chapter, we saw that communities also show variability through time because of the process of succession. As a result, the problem for conservation work is that, if we set aside a preserve to protect a shrubland community and in 15 years succession has made it a forest, have we succeeded? In other words, how can we protect a community when we know that it is naturally going to change over time?

Another major issue when working with communities is the general problem of habitat fragmentation (Figure 9.3). Human activities are drastically altering the structure of our landscapes, and as a result, large continuous tracts of habitat are being broken up into smaller and smaller isolated fragments. Ecologists have found that smaller fragments of habitat support fewer species than large fragments, partly as a result of what are called *edge effects*.

We can perhaps best visualize edge effects by first imagining a large continuous area of pine forest. If this forest is heavily logged so that all that remains after the logging are a few small isolated patches of forest, then we would essentially be left with tiny remnant fragments of pine habitat. If you were to stand at the edge of one of these fragments and compare conditions in the forest with the clear-cut area, how do you think they would be different? In general, outside the forest fragment, in the midst of the clear cut, there is more sunlight, higher temperatures, more wind, and probably lower humidity than inside the original forest. The edge of the forest experiences these changes more than the interior of the fragment, and these types of environmental alterations are referred to as edge effects. Edge effects generally make the fragmented habitat less attractive and less hospitable to plants and animals that normally live in the deep interior of a forest.

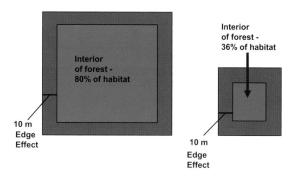

FIGURE 9.4. The edge effect has a larger proportional impact on smaller fragments of habitat than on larger fragments. If edge effects can be measured 10 meters in from the edge of a parcel of habitat, then a small fragment of habitat may contain very little habitat that is not impacted. (Figure by M. P. Marchetti.)

Edge effects also increase as the fragment size decreases (Figure 9.4), which means that small fragments are worse for habitat protection than large fragments.

In light of all these difficulties, if we want to protect a habitat or community, we have to first realize that not all parts of the habitat are the same and that not all communities are equal. As a result, protection efforts need to address the difficult question of "Which part of a habitat or area of a community is the best to preserve?"

SPECIES-AREA RELATIONSHIP

In order to answer the question above, we will turn to an area of ecology dealing with islands and biodiversity. A young E.O. Wilson and his student Robert MacArthur proposed some interesting ideas in 1967 regarding biodiversity and island ecosystems. Their ideas dealt with colonization and extinction rates on islands and were described in a book *The Theory of Island Biogeography*. There are two major take-home points from their work that we want to highlight: (1) larger islands tend to have more species than smaller islands, and (2) islands close to the mainland tend to have more species than islands farther away. If we look at a cartoon showing four possible islands (large/small and close/far) (Figure 9.5), which island do you think is likely to have the most species? Logic dictates that island C would have the most because it is the largest and closest to the mainland and that island B would have the fewest because it is the smallest and farthest away from the mainland. So how does

this theory apply to practical conservation questions about reserves and other protected areas?

The answer lies in our previous discussion on habitat fragmentation. We mentioned that human actions are fragmenting habitats and communities all across the planet. We are taking large tracts of continuous habitat and cutting them up into little remnants. We can envision these little bits of fragmented habitat that remain as "islands" in a "sea" of degraded habitat (Figure 9.6). These terrestrial fragments are ecologically similar in many respects to oceanic islands. They have discreet boundaries, are surrounded by a very different habitat, and may be hard to locate and colonize if you are a small individual. When we think of habitat remnants this way, it is easy to see how we could borrow the ideas of island biogeography for conservation science.

RESERVE DESIGN

Based on what we learned above and what we know from our study of ecology, we can begin to address one of the larger questions that first faced conservation scientists as the profession got its start: namely, How should we design the best possible reserve? This is also clearly bringing us back into the realm of in-situ conservation that we addressed earlier, although in this context we are dealing with the preservation of communities instead of individual species. When designing a community reserve, our goal is to include as many species as possible so we can protect the maximum amount of biodiversity.

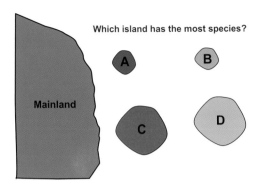

FIGURE 9.5. Island biogeography theory deals with the number of species found on islands. This number varies depending on the size of the island (large islands contain more species) and on its proximity to the nearest mainland habitat (near islands contain more species). (Figure by M. P. Marchetti.)

FIGURE 9.6. Habitat fragments, such as these forest plots in the midst of clear cuts, can act like habitat "islands" in a "sea" of non-habitat. (Photo from morguefile.com.)

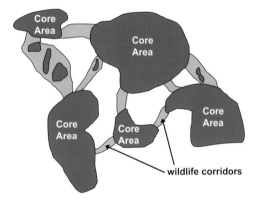

FIGURE 9.7. Habitat corridors are important conservation tools because they provide linkages between protected areas. Corridors have been shown to be used by animals such as bobcats and alligators and can even help in plant dispersal. (Figure by M. P. Marchetti.)

Given island biogeography theory, we want our reserve to have the largest possible area (i.e., to be a "large island") and to be as close as possible to an even larger area with the same or similar habitat (i.e., to be close to the "mainland"). Alternatively, if no pristine tracts of land are available, then our "mainland" will be any habitat that has not been heavily degraded by human action. In addition, we want our reserve to have the smallest possible edge. If you remember a little bit of geometry, you know that the shape with the smallest perimeter (edge) is a circle. So our reserve should be as close to a circle in shape as possible. Within the circle, it is important to have large amounts of habitat diversity and variability in terrain, in order to include the maximum number of species for protection.

All of the "rules" above provide a good solid start to designing a reserve, but with a few other additions, we can improve on this scheme. The first addition would be to have multiple reserves near each other and then to link them together with *habitat corridors* (Figure 9.7). Corridors are strips or bands of protected habitat that connect two or more isolated protected areas. Habitat corridors between areas can effectively increase the total size of a protected area. It has been demonstrated repeatedly that at least some large vertebrate species will use corridors to move between protected areas. A recent radio-collar study by

Dr. Seth Riley at the National Park Service in Santa Monica, California, has shown that mountain lions *(Felis concolor),* which are notoriously shy of people, will cross under a major freeway outside of Los Angles if they are given an underpass that connects two areas of good habitat. In Riley's study, this corridor was a small tunnel under the freeway used by equestrians to move their horses around. The mountain lions found this connection and were actively using it to expand their hunting and foraging range.

Additionally, a carefully designed reserve is best if it has a *buffer zone* around it, to reduce impacts of developed areas around the reserve. One vision of this is to create a central core

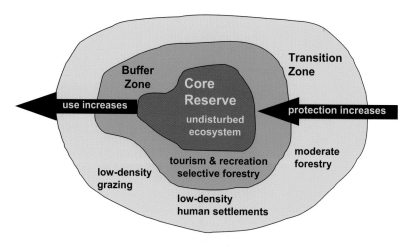

FIGURE 9.8. Buffer zones of differing use can act to insulate and protect sensitive areas and protected habitats. The ideal situation is to have a core protected area surrounded by a low-intensity use primary buffer zone, which in turn is surrounded by an additional higher-intensity use buffer zone. With this three-tier system, protection increases as you move toward the core area, whereas human use and impact decrease. (Figure by M. P. Marchetti.)

protected area, where protection is the strictest and human usage is the lowest (Figure 9.8). A core reserve would be the best way to protect animals and plants that are extremely sensitive to human contact. Outward from the core, human usage would increase, and protection would decrease. For example, a buffer zone around this core could be an area of intermediate use where human activities such as hunting, fishing, and hiking could occur. An additional buffer zone enclosing the whole region could be a zone of further protection, but one where more intensive management practices, such as sustainable forestry or low-density grazing, could occur. This type of buffering system can maximize the level of protection for the sensitive core area while also providing large areas for human use and occupation.

ECOSYSTEM-LEVEL CONSERVATION: WATERSHEDS

So far, we have talked about how conservation biologists go about protecting species and communities. Work at these levels has consumed much of the time and energy of conservation science to date, yet increasingly scientists recognize the need to expand the scope of conservation efforts to the level of entire landscapes and ecosystems. This is a relatively new and underdeveloped area of

research, so we decided to focus our attention here on one area within this burgeoning field, namely conservation at the level of the watershed or catchment.

Many discussions of preserves and protected areas focus on big blocks of a landscape or habitat region, yet flowing through these protected areas are streams and rivers that create a watershed. Watersheds are defined as the entire drainage basin of a given stream or river, from ridgetop to mouth. Because river systems are made up of small trickles feeding into brooks feeding into bigger streams feeding into rivers, watersheds are inherently nested within one another, much like the sets of Russian dolls that neatly fit inside one another. The boundaries between watersheds can be used to define landscapes and ecosystems and therefore can be approached as part of this larger ecological context.

Streams within a watershed may have their headwaters upstream of a protected area and their mouths far downstream. The linear nature of stream systems and watersheds therefore can pose difficulties for conservation work. For example, a logging operation in the headwaters of a stream may induce a landslide, creating sediment that flows downstream for many miles, into and through a protected area in the middle of the watershed. Alternatively, a dam downstream of a protected area may prevent migratory fish, such as salmon, from reaching their historic spawning grounds. Additionally a non-native species planted in a downstream reach of a river can invade upstream and become abundant in a preserve.

Not surprisingly, many protected areas in the United States and elsewhere have preserved beautiful forests or prairies with large populations of native mammals and birds and may simultaneously contain streams that are degraded or may harbor non-native species. For example, the Nature Conservancy has an amazing preserve along the McCloud River in northern California that contains old growth trees and a high diversity of terrestrial plants and animals. Yet the river is missing major players in its ecosystem. Sea-run Chinook salmon *(Oncorhynchus tshawytscha)* and steelhead trout *(Oncorhynchus mykiss)*, once present in large numbers, are now denied access to the river by Shasta Dam. Bull trout *(Salvelinus confluentus)*, once an important predator in the ecosystem, are now extinct. Brown trout *(Salmo trutta)*, an alien predator species, however, are common. Thus, the entire preserve has to be regarded as having lost significant aquatic biodiversity. It is likely that more has been lost from the watershed, because the salmon were once a major source of food for bears and other scavengers, which would carry the salmon nutrients into the surrounding forest ecosystems.

The fact that traditional parks and preserves often do not protect aquatic environments adequately is reflected in the growing realization that aquatic species and ecosystems are often the most endangered ecosystems worldwide. A recent evaluation of the status of species and ecosystems in the Sierra Nevada Mountains in California showed that aquatic systems were in the most trouble, with a suite of frogs, fish, and aquatic invertebrates being in danger of extinction. The best way to reverse this trend is to protect entire watersheds, using them to determine boundaries of protected areas.

To use another example from California, we can look at the challenges that arise when we try

FIGURE 9.9. Deer Creek Wilderness Area, Lassen National Forest, California. (Photo by M. P. Marchetti.)

to protect parts of the branched stream network that forms the upper Sacramento River watershed. At the highest point in the watershed, there are small unnamed tributaries that flow into Cub Creek (a rainbow trout stream). Cub Creek in turn flows into Deer Creek (a salmon spawning stream, Figure 9.9), which then flows into the mainstem Sacramento River. As we move from small to large, each of these nested tributaries has a progressively larger drainage area, making it increasingly difficult to manage as the size of the area increases.

For example, Cub Creek and all its tributaries are completely protected from logging and other insults because they are part of a U.S. Forest Service Research Natural Area. Cub Creek eventually flows into Deer Creek, which is of great conservation interest because it contains one of the last populations of wild spring-run Chinook salmon in California. Deer Creek flows through a designated wilderness area and parts of Lassen National Forest, but almost half of its watershed is owned by ranchers, timber companies, and small landowners. Protection of the Deer Creek watershed is being accomplished by a coalition of landowners (the Deer Creek Watershed Conservancy), working with state and federal agencies. The general framework of this protection stipulates that (1) private landowners must be allowed to continue to make a living from their land, (2) the national forest lands should be operated for overall public benefit, including extractive uses, (3) the key natural elements, such as Chinook salmon, must be protected and enhanced, and (4) actions taken in one part of the watershed are likely to affect all other parts of the watershed, so actions must be coordinated. The nature of this framework indicates that the Deer Creek watershed is not a preserve, in the traditional sense, but a managed ecosystem that includes humans as active participants in it.

Deer Creek eventually flows into the Sacramento River, the largest and perhaps most modified river in California. Because it provides much of the water for cities and farms in central and southern California, this river has been extensively dammed, diverted, channelized, polluted, and otherwise degraded. Yet there is a growing realization that even in this degraded ecosystem, the entire watershed must be managed in an integrated fashion, especially if the remaining natural features (such as the Chinook salmon and steelhead that must pass through the river to reach Deer Creek) are to be protected. An entire watershed management plan could ultimately benefit both the wildlife and the human inhabitants of the region by providing such amenities as a cleaner and more reliable water supply, increased protection from floods, and increased aesthetic and recreational value of the river and its tributaries.

The growing interest in watershed management results from a realization that (1) we all live in watersheds and (2) watersheds are a natural unit on the landscape, typically easy to define even if they cross political boundaries. In North America, citizen watershed groups are springing up like mushrooms after a rain because people are beginning to understand more clearly the enormous benefits to be gained from holistic watershed management approaches. Some of these benefits include

better protection for native ecosystems and their associated biodiversity.

LANDSCAPE-LEVEL CONSERVATION: ALASKA

Theory and just plain common sense indicate that there is no substitute for size when it comes to conservation: large natural areas are more likely to protect diverse communities of species and their ecosystems than are small ones. But large areas that can be managed mostly for their natural values are increasingly scarce. This is why Alaska and northern Canada are so important to North America and the world. Large tracts of relatively undisturbed lands and rivers still exist there. Some of these tracts are large enough to include some of the world's longest migrations of fish and wildlife, such as the caribou that migrate across hundreds of miles of tundra and the salmon that migrate hundreds of miles up the large rivers.

In Alaska, the modern conservation story began in 1959, when Alaska became a state despite its small population. A major issue for the statehood movement was federal control over the region's land and resources. Once statehood was granted, the federal government ceded large tracts of land (26 million acres) to the state, mainly to generate revenue for state operations. The Native Americans in Alaska, however, realized that their interests and those of the state of Alaska were often in conflict, especially after the discovery of oil in Prudhoe Bay in 1968. They therefore demanded land of their own, resulting in the Alaska Native Claims Settlement Act of 1971, which granted them control of 44 million acres, largely to generate revenue for the tribes. These major land transfers worried conservation organizations, which saw the need for formal protection for large areas.

In 1978, President Jimmy Carter declared 56 million acres of Alaska to be national monuments, and his Secretary of the Interior, Cecil Andrus, withdrew an additional 40 million acres from development. The U.S. Congress, in a 1980 compromise with the Alaska delegation, then passed the Alaska National Interest Lands Conservation Act (ANILCA), which was signed by Carter. This law protected over 100 million acres of federal land, most of it identical to that protected by Carter's own actions; this doubled the size of our national park and refuge system and tripled the amount of official wilderness areas. ANILCA expanded the national park system by over 43 million acres, creating 10 new national parks and increasing the size of three existing units.

What does all this mean for conservation? Most importantly, it means that large chunks of the landscape are protected from development with very limited human access, which is granted mainly to ecotourists, managers, and researchers. This, in turn, means that the animals that inhabit the region are freer to go about their business of life than they are in most parts of the world. The over 100,000 members of Porcupine caribou herd, for example, can continue to migrate over 640 kilometers (400 miles) from their winter range south of the Brooks Range and from areas in Canada's Yukon Territory, to their calving grounds on the National Arctic Refuge, and back again. Because no caribou walks in a straight line, the actual distance covered by each individual may exceed 4,800 kilometers (3,000 miles).

Likewise, sheefish, a large fish related to salmon and trout, should be able continue to make their annual late summer upstream migration from an estuary, hundreds of miles to spawning areas in the Kobuk River and similar large Arctic rivers. Their embryos incubate in the cold, calm water under the winter ice, and the hatchlings are washed downstream to the estuary for rearing. Unfortunately, to keep these types of large-scale processes going requires active management of the landscape, including the prevention of development and drilling for oil. Increasingly, it means figuring out ways to accommodate climate change, which is affecting Arctic areas more rapidly than other parts of the world.

Finally, it is worth noting that, in terms of area, the only larger region protected by the U.S. government is the 195,000 square miles of ocean surrounding some small U.S.-owned islands in the Pacific Ocean. In January 2009, President George W. Bush declared the region Northwestern Hawaiian Islands National Monument. Although remote, this region desperately needs protection from the effects of overfishing, which has changed the pelagic ecosystems in ways that affect everything from coral reefs to swordfish to monk seals.

CONCLUSION

Each species is a unique and separate natural entity that, once lost, can never be revived. The ideal goal for conservation biology would be to save all species, even every local population of species. Yet this is obviously not an option. When protecting species and establishing reserves, we are faced with difficult choices. Which species do we protect? Which habitats are most critical? How do we make these decisions in light of the stresses that overpopulation and human lifestyles place on these systems?

When considering which areas to conserve, we must weigh as many aspects as possible. Decisions must be based on a mixture of historical, evolutionary, community, and species approaches blended with a dose of reality. A World Resources Institute publication from 1989 recommended the following features to help evaluate the trade-offs and value judgments made in setting biodiversity and conservation priorities:

1. DISTINCTIVENESS Preserving an entire distinct species is clearly more important than saving populations of species with numerous representatives.

2. UTILITY When evaluating what to save, we clearly have to evaluate utility to humans of natural systems. For example, to humanity at large, large regions of tropical rainforest are extremely important, not only because they contain a variety of life, but also because they influence global ecosystem services such as climate. In contrast, an isolated, small woodlot in rural Indiana may have small global value, even if used by a few

migratory birds. Such a woodlot, however, may have an important local value (including education), which increases as it becomes part of a regional system of woodlots managed in part for conservation.

3. THREAT Saving the most beleaguered species and ecosystems must be done first. When establishing priorities, it is most important to focus on areas that are the most at risk. For example, Central America's tropical rainforests are less threatened than are the remaining fragments of tropical dry forest in that region, indicating that highest priority for conservation in the region should go to the latter.

Conservation biologists and society as a whole are faced with the problem that current rates of extinction far surpass those of most mass extinctions our planet has ever experienced. Under human domination, our planet is becoming a biologically impoverished relict of the world past generations have known. Already, we can no longer thrill to the sight of millions of migrating passenger pigeons darkening the sky, giant herds of bison thundering across the Great Plains, or the dayglow spectacle of breeding golden toads. We are a powerful biological force, and we are making choices that will influence the planet for centuries to come. In 50 million years, we humans may not exist. What species will exist will largely be a result of the actions we take today. What would a future

be like if all wildlife were evolved from rats, European starlings, and cockroaches? We are clearly the problem, but we can also be the solution, by using tools such as those discussed in this chapter. This is what conservation science is all about. It is a discipline focused on finding ways to make humanity more compatible with wildlife and wild ecosystems, using the best available science.

FURTHER READING

Groom, M. J., G. K. Meffe, and C. R. Carroll. 2005. Principles of conservation biology, third edition. Sinauer. Sunderland, MD. *This is the textbook that many biology majors use to learn more about the topics in this chapter. It is not as daunting as you might think and has a lot of great information.*

Quammen, D. 1996. Song of the dodo: island biogeography in an age of extinction. Scribner. New York. *Quammen is a gifted writer, and we would all do well to read and digest what he has to say in this book, which is written for a non-science audience.*

Rosenzweig, M. 2002. Win-win ecology. Oxford University Press. Oxford. *This is a book that is also somewhat hopeful in its message.*

Wilcove, D. S. 2008. No way home: the decline of the world's great animal migrations. Island Press. Washington DC. *A great book on one of the great conservation challenges of our time: protecting large-scale migrations of animals.*

Wilson, E. O. 2003. The future of life. Vintage. New York. *Wilson has written an impassioned plea to save the Earth's biodiversity. This is wonderful, if somewhat depressing, reading and is clearly worth the effort.*

10

Conservation and the American Legal System

As we have seen so far, during the last 150 years, conservation science has greatly expanded its scientific scope and rigor. In addition to its scientific achievements, conservation science has broken new cultural, social, philosophical, political, and legal ground. The most powerful and far reaching of these impacts comes from the legal backing that conservation science has been granted in the United States through the many federal and state laws that protect animals, plants, and habitats. These protective legal efforts did not originate de novo with the forefathers of conservation ideas, but instead arose gradually from a shift in political focus that began with the assignment of value to a narrow spectrum of charismatic vertebrates (such as the national symbol, the bald eagle). This movement eventually grew to encompass nearly all creatures and ecosystems on the planet, inspiring legislation in other countries. These political and legal changes reflect major shifts in knowledge and public attitudes that took place over a roughly 150 year period. Thus, the protective legislation inspired by the wanton destruction of wildlife in the nineteenth century laid the foundation for a long line of legislative efforts that eventually led to modern environmental legislation in the United States (see the table below), which includes the Endangered Species Act of 1973, perhaps the most stringent environmental law ever written (and enforced) anywhere on the planet. The baton for innovative and effective environmental legislation, however, has largely been passed on to other countries at present, especially those in the European Union.

In this chapter, we focus on some of the major federal legal actions in the United States (Table 10.1) as examples of laws that provide a broad spectrum of protection. Arguably, they provide the most extensive legal protections for biodiversity and wild lands of any country in the world, although, in practice, these protections are often overwhelmed by economic and social forces.

EARLY ENVIRONMENTAL LEGISLATION

The first legislative attempts to protect wildlife and natural resources began in the late nineteenth century, when it became apparent that relentless exploitation of natural resources was eliminating valued wildlife and forests. The greater Yellowstone ecosystem was the first major wild area to be

TABLE 10.1
Major Federal Environmental Legislation since 1872

1872	Yellowstone National Park founded
1897	Forest Management Act (aka Organic Act)
1900	Lacey Act
1903	Pelican Island Wildlife Refuge established
1906	Preservation of American Antiquities Act
1911	Fur Seal Treaty
1913	Weeks-McLean Act
1916	Migratory Bird Treaty Act
1916	National Parks Act
1934	Duck Stamp Act
1937	Pittman-Robertson Act
1950	Dingell-Johnson Act
1955	Clean Air Act
1964	Wilderness Act
1966, 1969	Endangered Species Acts
1969	National Environmental Policy Act
1972	Marine Mammal Protection Act
1972	Clean Water Act
1973	Endangered Species Act
1973	Convention on International Trade in Endangered Species
1976	Land Policy and Management Act
1976	Fisheries Conservation and Management Act (Magnuson Act)
1976	National Forest Management Act
1980	Comprehensive Environmental Response, Compensation, and Liability Act (Superfund)
1994	California Desert Protection Act
2009	Omnibus Public Lands Management Act

protected by the federal government, in large part because it seemed to be a piece of the "Old West" that somehow survived and was undeniably visually spectacular. Yet even Yellowstone had to be protected from exploitation in its early years by the U.S. Army. The first forest reserves were set aside, on paper, by the Forest Management Act of 1897 (aka the Organic Act), which led eventually (through additional legislation) to the founding of the national forest system.

THE LACEY ACT

The Lacey Act, sponsored by Representative John F. Lacey of Iowa, was passed by Congress in 1900 as a direct response to public outcry over the slaughter of herons and other migratory birds to furnish plumes for the garment and fashion industry, and in particular for ladies' hats. The Lacey Act made interstate transportation of illegally killed game animals a federal offense. This effectively curbed commercial hunting because the market for the plumes was in the northeastern states whereas the exploited birds lived in the southern states. The Lacey Act also sought to limit importation of exotic wildlife, such as the house sparrow and the mongoose, the introductions of which were detrimental to native wildlife. The Lacy Act was significant because it was the first law to bring the federal government unequivocally into the business of protecting wildlife. Because of its wide scope, the Lacey Act was supported both by recreational hunters and by the increasing number of bird protectionists who were organizing across the country.

PELICAN ISLAND

As a direct result of the campaign against plume hunting, President Theodore Roosevelt issued an executive order in 1903 declaring a small nesting area in Florida (Pelican Island) as the first federal bird sanctuary. Roosevelt's order unleashed a flood of nominations for other sites to be protected, and by the end of his first term, in 1904, 51 wildlife refuges had been created. In 1906, Congress confirmed the President's authority to set aside areas to protect wildlife and at the same time passed the Preservation of American Antiquities Act, which gave the President power to create national monuments. Today, the National Wildlife Refuge System, originated by Roosevelt, includes more than 540 sites, encompassing over 96 million acres in all 50 states and five trust territories. System-wide, the National Wildlife Refuge System receives almost 40 million visitors a year, but its real importance lies in the diverse wildlife and fish populations that it protects. In California, for example, there are 39 wildlife refuges, which range from "standard" refuges that protect waterfowl (largely to maintain populations for hunting) to those set up to protect seabirds and endangered species. Even refuges managed intensively for narrow purposes, such as hunting, provide habitat for a high diversity of other wildlife.

WATERFOWL PROTECTION

In the nineteenth and early twentieth centuries, market and commercial hunting (large-scale shooting of animals for sale) caused major declines in populations of wild ducks and geese. Sport and recreational hunters became alarmed at this and began pressing for conservation measures. Federal protection for waterfowl was created through a series of legislative acts that began in 1913 with the Weeks-McLean Act, which established federal control over migratory birds and ended springtime waterfowl hunting. This was an important step in protecting birds that were breeding or nesting in the spring season,

a feature of their natural history that makes them particularly vulnerable to overhunting. The Act allowed the Secretary of Agriculture to set closed seasons when it would be illegal to capture or kill migratory birds. Passage of the Migratory Bird Treaty Act in 1916, signed by the United States and Great Britain, established formal cooperation between the United States and Canada for the protection of both game and non-game birds. In 1937 the Migratory Bird Treaty Act was amended to include Mexico. Although this treaty was originally designed to protect only waterfowl for sport hunting, a major result was umbrella protection for all other migratory birds.

In 1934 the Duck Stamp Act, passed at the request of duck hunters, required all waterfowl hunters to purchase an annual federal hunting stamp. The first duck stamps were sold at $1 each, and a total of 635,000 were sold. Duck stamps today cost $10 each and are bought by stamp collectors as well as hunters. Proceeds from duck stamp sales are used to protect critical wetlands for breeding, migration stopover, and wintering of waterfowl. A significant aspect of the Duck Stamp Act is that it demonstrated the political and financial power of hunters and the growing recognition that habitats, not just wildlife itself, needed protection. The first duck stamp was drawn by J. "Ding" Darling, a political cartoonist who frequently focused on environmental issues (Figure 10.1).

Although commercial hunting was devastating to many species, sport and recreational hunting rarely endanger species and, in fact, provide a good source of revenue for wildlife protection and habitat improvement, as exemplified by the Pittman-Robertson Act. The passage in 1937 of the Pittman-Robertson Act, also known as the Federal Aid in Wildlife Restoration Act, created an additional major source of funds for wildlife protection by placing a 10 percent tax on the manufacture of sporting arms and ammunition. This act also stipulated that for states to receive P-R funds, all money they raised from the sale of hunting licenses had to be used for

FIGURE 10.1. Cartoon by "Ding" Darling railing against the impacts of hunting on waterfowl populations. Darling was a leader in the early conservation movement and helped to establish many wildlife refuges. (Reprinted with permission courtesy of the "Ding" Darling Wildlife Society.)

© 1999 J.N. "Ding" Darling Foundation

wildlife projects. Prior to the Pittman-Robertson Act, money intended for wildlife conservation often got redirected to fund other local projects such as schools or road repairs.

When federal funds first became available to the states, much of the attention of state-run wildlife agencies was concentrated on deer, elk, and other ungulates, the populations of which had been decimated during the days of market hunting. In the first 10 years following the Pittman-Robertson Act, 38 states acquired nearly 900,000 acres of refuges and wildlife management areas. Extensive replanting of trees and grasses was undertaken, and massive restocking efforts transplanted deer, pronghorn, elk, mountain goats, and mountain sheep to these restored areas. The Pittman-Robertson Act was so successful that 13 years later the Dingell-Johnson Act of 1950 was modeled after it to generate similar funds for fisheries projects.

THE FUR SEAL TREATY OF 1911

The northern fur seal was nearly exterminated by a century and a half of seal hunting on the Pribilof Islands near Alaska by Russian, American, Canadian, and British sealers. After the United States purchased Alaska and the Pribilofs from Russia in 1867, there was an attempt to limit harvesting of fur seals, but there was no regulation on seal hunting in the open seas. In 1911, the Fur Seal Treaty was signed by the United States, Russia, Japan, and Great Britain. The treaty provided protection for fur seals and sea otters on the high seas (outside the 3 mile limit that most countries claimed as territorial waters) from the four signatory nations. Each of the countries agreed to prohibit open ocean hunting and to manage populations of fur seals within its own territorial waters.

The treaty was upheld until the beginning of World War II, when it was terminated by Japan's entry into the war. From 1942 until 1957, the Pribilof fur seals were protected under a provisional agreement between the United States and Canada. In 1957, the North Pacific Fur Seal Convention was signed by Canada, Japan, Russia, and the United States. This interim treaty, which was similar to the 1911 Fur Seal Treaty, was followed by the Interim Convention on Conservation of North Pacific Fur Seals signed in 1963. These treaties called for intensified research programs and the establishment of a North Pacific Fur Seal Commission to study and make recommendations for management procedures and harvest quotas.

Fur seal populations have benefited from the protection and management provided by these treaties. Up from a low of about 125,000 in 1911, the current population is close to its former size of an estimated two to three million animals. Today, less than 2,000 animals are harvested each year, and the harvest is done entirely by native peoples for subsistence purposes. However, the populations of seals are now threatened again, this time by loss of their food supply, because heavy commercial fishing pressure has depleted the fish resources they depend upon.

THE ENDANGERED SPECIES ACTS OF 1966 AND 1969

Although legislation designed to protect birds and game animals has a long history in the United States, no legislation specifically focused on species threatened with extinction existed until the Endangered Species Act of 1966. This Act was fairly innocuous at its inception because its only provision was to authorize the Secretary of the Interior to maintain a list of rare and endangered species. All it did was create a list; it offered no legal protection or federal aid. In 1969, the Endangered Species Conservation Act expanded the listing process of the original

version slightly and authorized a prohibition on importing endangered species or products derived from them. In addition, foreign species under threat were allowed to be added to the U.S. list. Despite its weak beginnings, the establishment of the first Endangered Species Act was a major, if wobbly, landmark in wildlife protection because the legislation was created specifically for the benefit of wild animals rather than for direct human benefit (i.e., for maintaining wildlife populations for hunting or market use).

THE 1973 ENDANGERED SPECIES ACT

The Endangered Species Act (ESA) of 1973 was passed and signed by President Richard Nixon. It replaced the two previous endangered species acts with a comprehensive policy that went far beyond the earlier versions. Key provisions included (1) federal protection for any endangered organisms in the United States, including plants, invertebrates and vertebrates (provided they were recognized as a species, subspecies, or distinct population segment); (2) designation of habitat critical to the survival of a species; and (3) establishment of a procedure to determine the status of endangered species.

In contrast to earlier legislation that provided protection to endangered species only on federal land, the 1973 Endangered Species Act prohibited any taking of endangered species. The idea of "taking" is defined very broadly under the ESA to mean killing, harming, collecting, trapping, confining, harassing, and a whole range of similar activities. The 1973 ESA also extends protection to "almost endangered" species, now referred to as "threatened." Under this law, individual states are permitted to adopt more restrictive legislation.

The 1973 ESA explicitly recognizes species as integral components of ecosystems and stresses that the integrity of ecosystems must be maintained. The stated purpose of the Act is to protect and enhance populations of endangered species through a variety of conservation steps. These steps include (1) official listing of species as threatened or endangered, (2) prohibition against taking of listed species, (3) acquisition of habitat for listed species, and (4) development and implementation of recovery plans for listed species. Unfortunately, each of these steps comes with a variety of stumbling blocks that makes the reality of endangered species protection quite different from the theory and philosophy behind the act.

According to the 1973 ESA, the Secretary of the Interior is required to establish a list of threatened or endangered species. The criteria for endangerment may include any reason, natural or human caused, for the decline of a species or population. The official status of a species that has undergone a population decline can be divided into three overlapping categories: "endangered," "threatened," and "species of special concern." *Endangered species* are those whose populations have declined to a point where extinction is imminent if action is not taken to protect the species. *Threatened species* are those whose numbers are declining and are likely to become endangered in the near future if protective action is not taken. *Species of special concern* are those showing decline, or which have a very limited range, but which are not known to be faced with extinction in the immediate future. One thing you may notice about the above list is the slightly vague terminology used in defining the categories. In practice, it is not always easy to demonstrate that a species is threatened with "imminent" extinction.

You may be wondering who gets to list a species as threatened or endangered. The process of listing a species on the federal endangered species list may be initiated by any individual, agency, or group that submits a petition to the Secretary of the Interior. The weight of the supporting evidence supplied by the petitioner is judged by the U.S. Fish and Wildlife Service (USFWS) or, if a marine or anadromous species is involved, by the National Marine Fisheries Service (NMFS). Currently, there are over 1,300 species listed as threatened or endangered. The NMFS has

regulatory authority over 65, and USFWS has authority over the rest. Some of the criteria used to determine whether a species should be listed as threatened or endangered include (1) habitat destruction, (2) overexploitation, (3) potential eradication due to disease or predation, and (4) inadequate protective regulations.

The formal listing process has four steps. First, a species is nominated for review of its status, through the submission of the formal petition or white paper. The second step is an evaluation of the petition to determine whether it warrants further review. The nomination is then either rejected (which rarely occurs) or is accepted for a second review. This second review is a holding step during which more ecological and conservation information is collected and evaluated. This step is often used in controversial cases to slow down the listing process, especially if funding is lacking for studies. Third, the USFWS staff solicits data on the species and prepares a status report to determine whether a species should be listed, or given a change in status. This report is then evaluated by USFWS and must be approved by each local, regional, and national office before further action is taken. The final step in the listing process is a notice of proposed listing, which is published in the Federal Register. At least 60 days must be provided for public comment. During this time, outside parties (people or agencies with political or economic interests for or against the listing) can request a public hearing on the proposal. If no further review is required, the species then receives its final listing in the Federal Register, at which point the species is granted full protection under federal law.

In theory, listing a species should require 195 to 255 days between the time a petition is received and its final listing (or rejection) in the Federal Register. In practice, however, the listing process typically takes at least two years and sometimes significantly longer. For many species, the process is tragically hindered because the understaffed federal agencies

TABLE 10.2

Number of species listed under federal ESA as of 2008, and percentages for which recovery plans exist.

GROUP	ENDANGERED	THREATENED	RECOVERY PLAN
Mammals	69	12	69%
Birds	75	15	94%
Reptiles	13	24	100%
Amphibians	13	10	74%
Fishes	74	65	73%
Insects	47	10	61%
Total animals	448	160	81%
Total plants	598	146	89%
Total	1,046	306	86%

reviewing petitions are forced to file reports saying they cannot list a species because they have inadequate time or information to do a proper evaluation.

In addition to listing, the 1973 Endangered Species Act mandates that a recovery plan be developed for all species that are listed. Recovery plans are intended to identify the causes of a species' decline and to specify actions needed to reverse the decline, including the designation of critical habitats necessary to facilitate their recovery. Ultimately, the goal of the ESA is to remove species from the endangered species list because they are no longer in danger of extinction.

LISTED SPECIES

It is interesting to examine some of the details of the species that have made it onto the list. As of 2008, 1,352 species have been listed in the United States as either threatened or endangered; of these, 608 are animals, and 744 are plants (Table 10.2). For a fairly large percentage of these (14 percent), there are currently no recovery plans in place. In addition, only 506 of these listed species have designated critical habitat. In terms of the number of species waiting for listing, 143 animal and 138 plant species are currently candidates for inclusion.

Other interesting patterns emerge when we look at the number of listed species by state (Figure 10.2). Here, we can see that disproportionate numbers of threatened and endangered species are found in island states (Hawaii and Puerto Rico); in large, dry states (California and Texas); and in southern states (Florida and Alabama). The numbers of threatened and endangered species in states with these characteristics serve to illustrate what we learned earlier in this text about patterns in global biodiversity: namely, that there is more biodiversity closer to the equator.

POLITICS AND LISTING

Interest groups play an important role in the implementation of the ESA. It has been estimated that, since 1973, at least half of the listings have resulted from the presence of a visible constituency or advocacy group. Pressure from interest groups has a strong impact on the action taken by the listing agency. In general, if a species is to be put on the protected list, it has to have an advocate or champion within the USFWS or in an environmental or scientific group. When an effective advocacy group lobbies for a species, it often speeds up the listing process and may also result in a higher degree of protection for a species than would otherwise have been provided.

The decision to list (or not list) a species is dependent upon a huge amount of administrative discretion. Although the 1973 ESA states

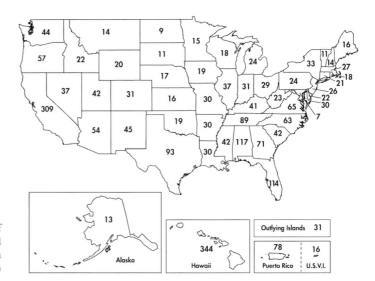

FIGURE 10.2. Number of federally listed endangered species by state, 2008. (Drawn using data from USFWS.)

that the values of species to be protected include aesthetic, educational, historical, recreational, and scientific, it is generally scientific and commercial data that provide the best argument for legal protection of a species. Mathematical and population models, based upon ecological data, are often used to predict the future population trends of a species. However, for most endangered species, neither historic nor current data actually exist to adequately describe the basic population ecology of the species, making this task difficult. Furthermore, an understanding of the species' role in its ecosystem and its interactions with other species is also frequently sketchy, and the impact of human-induced change on the species is often unclear. Consequently, many listing decisions must be made on the basis of woefully inadequate information. Even seemingly straightforward taxonomic questions such as "How do we define a species or sub-species?" can lead to problems. Grizzly bears, for example, have had subspecies classification difficulties several times over the last 20 years. The decision to protect an individual population of grizzly bears depends upon whether one defines 86 different grizzly bear subspecies, or just two.

The many problems and challenges with the ESA have led to considerable controversy in its reauthorization. Indeed, as of 2008,

reauthorization of the ESA has not been accomplished. The act formally expired in 1992 but has been kept alive by annual congressional appropriations for its continuance. Public opinion polls indicate that the ESA is very popular with the American people, who appreciate its success at protecting symbols of the wild such as the bald eagle.

However the act is not without its critics. Even among supporters, there are some ecologists and environmental groups that have difficulties with it, for reasons including the following: (1) its focus on species makes it too narrow when preservation problems are so large, (2) a more ecosystem-based approach would be preferred, (3) the act is more reactive than proactive in its protection, and (4) the focus on individual species on the brink of extinction often leads to costly and drawn-out legal battles. On the detractors' side, some politicians and developers attack the act as interfering with private property rights and economic activity, although such allegations are usually exaggerated.

The successes of the ESA are not as great as some had initially hoped. As of 2007, only 41 species have been de-listed, 16 due to recovery, nine due to extinction, nine due to changes in taxonomic classification, five due to discovery of new populations, one due to an error in the listing rule, and one due to an amendment to the Endangered Species Act. Twenty-three others have been downlisted from "endangered" to "threatened" status. Yet the majority of the success stories of the ESA are often overlooked by the general public. Since being listed as threatened or endangered, 93 percent of all species have had their population sizes increase or remain stable, which in and of itself is a great measure of success. For example, whooping crane numbers have gone from a low of 56 individuals when the species was listed to 355 birds in the wild in 2007 and about 150 in captivity. Likewise, the population of a small species of fish called the Big Bend Gambusia went from a few dozen individuals to a population of over 50,000 following protection. In this way, the major success of the ESA has been the immediate protection of species from extinction, rather than removal from the list entirely. With increased federal funding and continued interest by the general public, success stories of the ESA will undoubtedly continue.

HABITAT CONSERVATION PLANS

Because the ESA was perceived as seriously interfering with development forces and the use of private lands, in 1982 it was amended to allow the development habitat conservation plans (HCPs). HCPs, in essence, are agreements between the federal government and private landowners that allow a landowner to destroy or alter the critical habitat of an endangered species in exchange for protecting similar or better habitat elsewhere, or by contributing to a program to restore degraded habitat. HCPs are being developed at a rapid rate, especially in California. There is widespread agreement that the concept of HCPs is good, in that they focus on protecting habitats and can defuse long and expensive court proceedings. They are nevertheless highly controversial for the following reasons:

- They require that long-term decisions on habitat protection be made using currently available information and data, which are usually extremely limited. HCPs typically have a timespan that lasts 50–100 years, meaning that the decisions made with poor information today can be in place for generations to come.

- They contain a "No Surprises" clause, which says that if new information or new endangered species are discovered on private land covered by the HCP, the landowner is exempt from doing anything about it. This means that if critical habitat for an endangered species is discovered on private land covered by an HCP, the federal government will have to work out a protection agreement with the land owner or buy the land if it wants to save the habitat and the species.

- Most are covered by a "Safe Harbor" policy, which establishes a baseline level for the population of an endangered species on land covered by the HCP. If the population numbers rise above the baseline (e.g., by immigration from yet-to-be-developed land), the landowner can remove the "extra" individuals (by transplanting them to another area) without fear of penalty. The purpose of this policy is to eliminate the motivation of landowners to destroy habitats on undeveloped land for fear it will be colonized by an endangered species.

- HCPs are very expensive and time consuming to set up and monitor. The U.S. Fish and Wildlife Service, the main agency responsible for establishing, monitoring, and enforcing provisions of HCPs, is a relatively small agency that does not have the resources for this task.

As you can imagine, HCPs are contentious and often generate a lot of debate. Perhaps the most ambitious HCP to date has been established to protect the endangered California gnatcatcher (a bird) and other plants and animals found mainly in coastal sage scrub habitat in southern California. The land this community occupies encompasses areas of five southern California counties, some of the most valuable real estate in the world. Most of it has been slated for development to accommodate the skyrocketing human population of the region. The HCP in this case is being developed through a similar process established by the state under its own endangered species act. This process has defused the major ESA fight looming between conservationists and developers (although it is highly contentious itself) and is providing some protection to lands that would otherwise be developed. Whether the coastal sage scrub community and its indigenous flora and fauna will survive the process is not known, but many regard the likelihood of survival as being much higher with the HCP process than without it.

NATIONAL ENVIRONMENTAL POLICY ACT

In 1969 Congress passed the National Environmental Policy Act (NEPA), which requires environmental impact statements (EIS) to be written for all federally funded projects. NEPA also directs all federal agencies to prepare an EIS for any federal action that may significantly affect the quality of the human environment. An environmental impact statement is a document that identifies, predicts, evaluates, and proposes mitigation for the biological, social, and other effects of a development project before major decisions or commitments have been made. In effect, the process ensures that all projects using federal funds must consider all environmental impacts and must make this information available to the public in the form of a written environmental impact report (i.e., an EIS). NEPA was designed to ensure that pertinent information about potential adverse effects of development projects would be available to the general public before a project began in earnest.

In a very real way, this legislation marked the beginning of public participation in environmental decision making and provided the impetus for much environmental litigation. Ideally, NEPA is intended not only to require evaluation of potential projects but also to encourage critical evaluation of the environmental effects

of possible alternatives. Actual implementation of NEPA has resulted in numerous court cases concerning many aspects of development plans. In the first five years of NEPA, federal agencies filed nearly 7,000 draft environmental impact statements. The number of preliminary reports (environmental impact assessments, or EIAs) used to determine the need for the more elaborate analysis of an EIS was many times greater. For example, the Army Corps of Engineers produced approximately 10,000 EIAs in 1975, resulting in 273 environmental impact statements.

When the NEPA process is working well, several alternative courses of action are spelled out, and the one least harmful to wildlife is chosen for implementation. Given the economic expectations of many projects subjected to the NEPA process, the option chosen is usually not better than the status quo (e.g., wild lands supporting wildlife), but it does involve protecting some habitat or mitigating for habitat lost.

MARINE MAMMAL PROTECTION ACT

The Marine Mammal Protection Act (MMPA) of 1972 was the first national legislation to emphasize the stability of the ecosystem as the primary objective of management. This new law also established an indefinite moratorium on the killing or harming of all species of marine mammals, pending review of the status of each by the Secretary of the Interior. Additionally, the MMPA has established cooperative state-federal research programs with grants to organizations and universities for research into the population status, limiting factors, and general health of all marine mammal populations. Eskimos, Aleuts, and Native Americans are exempt from the take ban and are allowed to kill animals for meat, ivory, and hides for traditional cultural uses.

With the Marine Mammal Protection Act, jurisdiction over marine mammals is split between the Commerce and Interior departments. The MMPA also contains provisions for a "depleted" category, which is similar to the "threatened" category in the Endangered Species Act. Legislation such as the MMPA, NEPA, and

FIGURE 10.3. Harbor seal (*Phoca vitulina*), a species protected under the MMPA. (Photo by Greg Cunningham.)

ESA can be seen as efforts to begin the process of conferring basic rights on nonhuman organisms in the natural environment. Additionally, passage of the MMPA has provided further international thrust to species protection.

Since the passage of this act, populations of marine mammals have increased in abundance, to the delight of the general public (Figure 10.3). However, fishermen and others are complaining that sea lions and orcas can also prey on endangered salmon, leading to confrontations over enforcement of the act. For example, when sea lions were observed feeding on endangered steelhead trout in the Ballard Locks in Washington, 29 individuals were captured and relocated long distances from the locks; most were back in a few days. The basic problem is that dams and other structures that restrict movement or concentrate fish are magnets for sea lions, which find the fish to be easy pickings. Efforts to harass the sea lions away from the endangered fish have largely failed. The MMPA allows killing of problem animals in such situations, but such action has not yet been implemented.

CONVENTION ON INTERNATIONAL TRADE IN ENDANGERED SPECIES

The Lacey Act of 1900 was the first legislation to restrict international animal trade. However, such activities continued to pose a major threat to wildlife for many decades. First drafted in the mid-1960s, the Convention on International Trade in Endangered Species (CITES) was signed by 87 nations in March 1973. The goal of CITES is to protect the viability and survival of species from threats caused by international trade involving those species or products made from them. Each member nation is required to establish its own agencies to regulate the import and export of endangered or threatened species and to appoint a scientific advisory authority. CITES also affects countries that did not choose to sign the treaty, by eliminating the market for illegally obtained species. The signing of CITES also pressured the United States to establish stronger domestic legislation to set an example for other countries and to develop a protocol for implementing such regulations.

CITES is currently signed by more than 170 countries, and representatives from member countries meet periodically to decide which species should be included in the trade bans. Species that are included are placed in one of two categories, Appendix I and Appendix II. Appendix I species are considered endangered, and commercial trade is not permitted. Both import and export permits are required at every border. Appendix II species are threatened and could become endangered if trade is not regulated. Export permits are required. Roughly 5,000 species of animals and 28,000 species of plants are protected by CITES against overexploitation through international trade, including species such as the black rhino, which is often killed solely for its horn (Figure 10.4). The endangered species are placed in the appendices, according to how threatened they are by international trade.

One of the biggest strengths of CITES is that the burden of proof is placed on the exploiter. In other words, CITES requires potential exporters to prove that a species is not on one of the lists before trade in that species will be allowed. This philosophical shift is in stark contrast to the typical "innocent until proven guilty" philosophy used throughout most of our legal system. CITES, in a sense, forces the exporting

FIGURE 10.4. Black rhino (*Diceros bicornis*) in Etosha National Park, Namibia. Rhinos are of particular concern for conservationists because their population numbers are small and their horn is prized for use in traditional Chinese medicine. (Photo by M. P. Marchetti.)

organization to prove that it is not guilty. CITES has allowed the fate of a number of high-profile endangered species to be played out on the international arena; the best example of this is the African elephant *(Loxodonta africana)* and trade in elephant ivory.

CITES AND IVORY

Estimates of African elephant numbers in 2006 were as low as 10,000 animals total, down from over 400,000 in 1970. To satisfy world demand for ivory in the past, an estimated 200 to 300 African elephants were killed every day. A quota system for legal, regulated trade in ivory was authorized in the 1970s by CITES. However, because at least 80 percent of the world's ivory trade is believed to occur illegally, the quota system was ineffective. Obviously, the best way to halt the destruction of the elephant populations is to halt the demand for ivory products.

In June 1989, the United States banned the import of both raw and worked ivory from all countries. This ban is aimed at individual souvenirs purchased abroad by tourists as well as large commercial imports. It will remain in effect at least until adequate sustainable harvest levels are determined and enforceable international ivory trade controls are established. Up until the recent importation ban, the United States imported 10 to 12 percent of Africa's annual ivory export for piano keys, jewelry, and trinkets. About 65 percent of the U.S. imports came from Hong Kong, which is the major world ivory dealer and carver. Both the United States and Hong Kong are members of CITES. Japan, another member of CITES, is the world's leading consumer of ivory. In October 1989, CITES moved elephants from Appendix II to Appendix I, resulting in a complete ban on the trade in elephant products. The result was a spectacular 30 to 50 percent drop in the market value of ivory.

The U.S. ivory moratorium does not prevent sport hunters from importing trophies of African elephants legally taken in the country of origin, providing that the country has a CITES ivory export quota and has issued appropriate export permits. However, legal sport hunting has not been a significant factor in the elephants' decline. Furthermore, the license fees and other expenses associated with legal sport hunting contribute to wildlife management programs and give African countries an additional economic incentive to maintain huntable herds. Interestingly, South Africa and Zimbabwe both have large managed elephant populations and derive income for wildlife conservation and park management from the sale of ivory.

In recent years, CITES has lifted the ban on export of ivory from some African countries (e.g., Botswana, Namibia, Zimbabwe) that are presumed to have well-managed herds. The results have been mixed at best, in part because it is hard to distinguish legal ivory from that harvested illegally in other countries.

THE FUTURE

It is significant that the strongest environmental legislation stems from the 1970s and was signed into law by presidents Nixon and Ford. Since then, although strong legislation has been passed to reduce various kinds of pollution and contaminants in food, advances in environmental protection have largely come from minor presidential orders, small changes in existing regulations, or initiatives by states. If anything, the protection granted by the various environmental acts has actually been reduced by a succession of federal administrations that have failed to provide sufficient funding to support legislative provisions, appointed administrators unsympathetic to environmental protection, attempted to change provisions in existing environmental laws, failed to reform key harmful laws (e.g., the Mining Act of 1872), and chosen not to cooperate with global imperatives in environmental protection (e.g., the Kyoto Accords). It is to the credit of the American people that Congress and several presidential administrations did not succeed in their efforts to dismantle any of the major provisions of the Endangered Species Act, although few species have been listed in recent years, despite the obvious need to do so. Occasionally, protective legislation did get through the largely unfriendly Washington political process of the past decades, such as the California Desert Protection Act of 1994, which placed several million acres of desert habitat under the protection of the National Park Service (albeit with inadequate funds to manage it properly). The question remains as to whether all this legal protection is sufficient to protect biodiversity, given the rapid changes our planet is undergoing at the present time and our unwillingness to adequately enforce existing laws. The administration of President Barack Obama, fortunately, is working to reverse the decline in protection left by past administrations. One of his early actions (March 30, 2009) was to sign into law the Omnibus Public Lands Management Act, which protects 2 million additional acres as wilderness and increases the number of rivers designated as wild and scenic by 50 percent. In his remarks on signing the bill, President Obama said, "This legislation guarantees that we will not take our forests, rivers, oceans, national parks, monuments, and wilderness areas for granted, but rather we will set them aside and guard their sanctity for everyone to share. That's something all Americans can support."

CONCLUSION

The American people are justifiably proud of their record in protecting wild areas and species through governmental and legislative action. It is significant that there has been no new ground-breaking legislation, however, since the early 1970s, when Republican Richard Nixon was president. Despite the popularity of our parks, monuments, and refuges, and despite a public that consistently supports strong environmental laws, there has been a distinct downturn in protection of endangered species and habitats in the last decade. It is hard to be optimistic when we see enormous economic interests pitted against the underfunded agencies and private organizations dedicated to the salvation of species. Yet progress has been made, and at least a few species have been pulled back from the brink of extinction.

Such progress may be temporary, however, given the increase in human populations and resource use, and the accompanying global climate change. Ultimately we will need not only stronger laws but also a public dedicated to taking strong action to protect wildlife species and halt global climate change. Hopefully, in the coming years, the United States will finally move to not only ratify international global change accords, but also to assume a long awaited leadership role in dealing with this enormous threat to the planet's biodiversity. In the long run, species preservation will depend on us changing our lifestyles and making sure there continues to be room on this planet for more than just human beings. Laws are never enough by themselves.

FURTHER READING

Dombeck, M. P., C. A. Wood, and J. E. Williams. 2003. From conquest to conservation: our public lands legacy. Island Press. Washington, DC. *Drawing upon their vast experiences, the authors* bring a unique sense of insight into the internal conflicts, competing interests, and contradictions inherent in the federal agencies that managing our public lands.

National Research Council. 1995. Science and the Endangered Species Act. National Academy Press. Washington, DC. *A solid overview of what scientists know about extinction, and what this understanding* means for the implementation of the Endangered Species Act.

Tober, J. A. 1989. Wildlife and the public interest: nonprofit organizations and federal wildlife policy. Praeger Publishers. Westport, CT. *This book provides a sketch of the influential nonprofit wildlife organizations, the people they serve, and the methods they use to influence public policy.*

11

Invasive Species
and Conservation

The introduction of invasive alien (non-native) species into an area can result in a loss of biodiversity. This statement seems to contradict itself, but it is true. The reason is simply that when alien species invade an area they often drive native species to extinction. Multiple invasions of alien species thus paradoxically can *increase local biodiversity but decrease global biodiversity.* Unfortunately, everywhere humans settle, they bring along familiar animals and plants. As a result, a few hardy species are enjoying worldwide distributions whereas endemic native species (those that occur nowhere else) disappear. Species that are part of this increasingly worldwide fauna include Norway and black rats, the house mouse, the muskrat, the feral cat, the feral goat, the house sparrow, the European starling, the common carp, the brown trout, the rainbow trout, the mosquitofish, the Asian clam, the Argentine ant, the tiger mosquito, and the cabbage butterfly. In fact, alien species are consistently ranked second behind habitat change as a major cause of biodiversity loss. Typically, these two factors work together because altered habitats often favor alien invaders. This chapter deals with the process of species invasion and explains why species are introduced, why they are successful, and how they impact native species.

INVASIVE SPECIES

Alien, exotic, non-native, non-indigenous: all of these terms are adjectives used to describe essentially the same thing, *a species that has been moved by human activities to an area where it was historically absent and where it did not evolve.* Generally, humans help these species cross natural barriers, such as mountains and oceans, which had previously limited the species' movement. For instance, the many plants and animals that the colonists brought from Europe to North America crossed a major barrier, the Atlantic Ocean, with the help of sailing ships that carried people.

The number of species introduced worldwide is staggering and is growing rapidly due to the development of modern rapid transportation systems that move both humans and cargo over great distances very fast. Tropical fish collected in the wilds of South America can be sold in a U.S. pet store three days after they are caught.

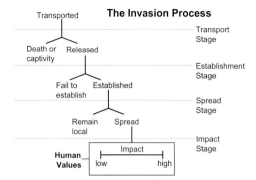

FIGURE 11.1. Species invasion is a process with distinct steps: transport, establishment, spread, and impact. For an invasive species to have an impact, it must proceed through all the preceding stages. Note that not all introduced species make it through all the stages and that conservation efforts aimed at reducing invasions are likely to be vastly different at each stage. (Reprinted with permission from Lockwood, J., M. F. Hoopes, and M. P. Marchetti. 2006. Invasion ecology. Blackwell Press. Oxford.)

Invertebrates sucked into the ballast water of ships can be dumped into an entirely different ocean after a ride of two weeks. An insect pest burrowed into a piece of fruit can be carried thousands of miles in a single airline flight.

It is important for us to realize that a species invasion is not a single event but a series of individual events all linked together (Figure 11.1). The first step in this invasion process is clearly *transportation*. Somehow (as we will see below), a species has to get moved to a new location, but just because it is picked up and moved doesn't mean that it will invade a new area. It is quite possible that the transported individual will die or be captured during the transport stage. If the species is successfully transported, it then needs to be released and to survive in the new area. This is called the *establishment stage,* and some species do not make it past this stage. For example, thousands of Chinook salmon from California were introduced into rivers of the East Coast in the nineteenth century, but the fish never became established despite all the effort. If a species manages to establish a population in a new area, the next stage is population growth and spread of the species *(spread stage)*. During all stages of the invasion process, human values (i.e., how we view the species) play an important role in the success of an alien species, but it is in determining the impact of an invasion *(impact stage)* that this becomes most noticeable. For example, we encourage species that have high economic value, even if they harm native species. Most of our major agricultural crops (tomatoes, apples, rice, etc.) were brought here from other regions of the world and occupy millions of acres that once supported native species, yet we do not consider them to have much of an impact because we value them as staple crops. On the other hand, species such as the zebra mussel *(Dreissena polymorpha)* are considered to be invasive pests with a very negative impact. The zebra mussel plugs up water intakes for cities, costing us billions of dollars annually and altering aquatic ecosystems throughout North America because it filters food organisms from the water. It is clear that human perception and the ways in which we value things play a large role in how we handle alien species.

WHY ARE SPECIES TRANSPORTED?

As humans have spread, so have many other species, taking advantage of the "new" environments created for them by us. The reasons for the transport and release of non-native species are as diverse as the plants and animals themselves but can be broken into two large categories: intentional and byproduct introductions.

INTENTIONAL INTRODUCTIONS

It is perhaps easiest to understand the intentional transport and introduction of species by humans, because this process has been occurring since ancient times. For example, ancient Chinese texts record plants and animals being introduced to new locations over 7,000 years ago. Because of the diversity of intentional introductions, it is useful to categorize the reasons for them into the broad groups below.

CULTURAL IMPERIALISM During the European settlement of the New World, there were many plants and animals brought over because the colonists were unfamiliar with New World species and thought the familiar species from back home were superior to the unfamiliar ones. This rather arrogant attitude has been called cultural imperialism, and it went hand in hand with the conviction that Europeans were superior to native peoples as well. As touched on in the chapter on the history of humans and wildlife, many European settlers feared the wildness of the Americas, so introducing species from their homelands was a means of reducing their uneasy feelings (the old version of homeland security). Even today, many of our most familiar animals and plants are European in origin, including the grasses along the road-sides, the house mice in our homes and gardens, and the horses in our pastures. The four species discussed in case studies toward the end of this chapter (the European starling, house sparrow, common carp, and brown trout) were all brought intentionally from Europe because they were believed to be superior to their local equivalents.

FOOD We humans like to eat certain plants and animals, and so we introduce them to be handy for harvest. Crayfish are a good example of this. Red swamp crayfish *(Procambarus clarki)*, native to Florida and the swamps of the southeastern United States, were introduced into California as a food source in the 1930s. They quickly escaped from their farm ponds and have now spread extensively across most of the western United States and even into Hawaii. Likewise, the large signal crayfish *(Pacifastacus leniusculus)* from the western United States was introduced into Europe with disastrous consequences for native crayfish. In a similar manner, bullfrogs *(Lithobates catesbeiana)*, a common invasive species in the western United States and other parts of the world, were introduced as a food source when French cooking and frog legs became a fad. Many people who eat these species would still regard their introductions with favor, despite the negative effects on the local fauna.

ECONOMICS During the European settlement of North America, animals such as muskrats, nutrias, beavers, and arctic foxes were introduced into new areas to support the fur trade. Even though there were positive economic aspects of these introductions, there were many negative ecological aspects. Introduced muskrats in California and nutria in Louisiana created major problems by burrowing into levees, causing them to collapse. Many plant introductions were also made for economic reasons, as the huge horticultural industry attests. Most common garden and landscaping plants are not native to wherever they are planted. For example, in the hope of providing better forage for cows, the fast-growing kudzu vine was introduced to the United States in the 1870s. Subsequently, this species has spread and has literally taken over many areas in the southern part of the United States because it grows extremely rapidly (up to 12 inches per day).

RECREATION Many familiar fishes, mammals, and birds around the world were introduced to provide opportunities for fishing and hunting. Examples include the ringnecked pheasant, wild boar, and common carp, as well as many species moved outside their native range, such as rainbow trout and largemouth bass. However, these species often end up doing harm because they threaten the existence of desirable native species. Recent studies in the Cosumnes River in central California have shown that the alien redeye bass *(Micropterus coosae)*, introduced as a game fish for small streams, has now almost completely displaced the native fish fauna in the river system. Because the bass is so small (a 25 cm fish would be a whopper), few people fish for it. Recreational fishermen also are responsible for the widespread introductions of bait minnows, released after the end of a day of fishing. In isolated lakes, such minnows can become abundant pests and can actually compete with gamefishes for food.

BIOLOGICAL CONTROL When a species, especially an alien species, becomes so abundant that it creates problems, other organisms are often introduced to control it. Unfortunately, control species often become pests in one area even if they are effective in their "job" somewhere else. A classic example of this is the introduction of the cane toad *(Bufo marinus)*, which was

FIGURE 11.2. The invasive cane toad *(Bufo marinus)* was introduced into Oahu, Hawaii, in 1932 and has subsequently spread to all the Hawaiian islands. (Photo by Julie Nelson.)

introduced to many islands in the Caribbean and Pacific to consume unwanted insect pest species (Figure 11.2). In some places, such as Puerto Rico, the cane toad may have been successful for controlling pests in sugar cane, but in most places, such as Australia, the introduction was a biological disaster, and the voracious toad threatens native biodiversity in the areas it has invaded. This is not to suggest that all biocontrol species have been failures; under some conditions, biological control may be a useful management technique for controlling a pest species. For example, mosquitofish *(Gambusia spp.)* are widely planted in rice fields and urban areas around the world to control mosquitoes and gnats. In these situations, they are an effective and inexpensive alternative to pesticides. In less disturbed systems, however, they can do damage to native fish, amphibians, and invertebrate populations. In general, intentional introductions of new species for biological control often have unanticipated consequences to ecosystems and so should only be done after a scientific validation study.

PETS The intentional release of pets is a surprisingly common method of introduction. Hundreds of alien species are brought to North America to occupy aquaria, cages, and backyards, and many are released by owners who grow tired of taking care of them. In the major-

ity of cases, these introduced pets die of stress or starvation, or are eaten by predators soon after their release, so they do not become a problem. However, there are numerous cases where the release of pets has resulted in the establishment of harmful populations.

There are many alien fishes that have been released in large numbers by hobbyists, which is one reason that goldfish *(Carassius auratus)* and guppies *(Poecilia reticulata)* have nearly worldwide distributions. Look in the urban ditches and ponds in tropical countries, and you will frequently see guppies, or their relatives the swordtails and mollies, swimming about. In colder areas, guppies thrive mainly in sewage treatment plants into which they have been flushed. Because many fish, like guppies, are tropical, they are only able to inhabit warm waters in North America; hence, introduced aquarium fish are especially common in the waters of Florida, Southern California, Nevada, Arizona, New Mexico, and Texas. In many cases, these species have become pests and are virtually impossible to eradicate. The combination of predation and competition from alien species has put many native species on the endangered list.

Alien birds have also found their way into areas of North America and Hawaii by means of unwise hobbyists. The intentional release of colorful tropical birds such as parrots and parakeets has been quite common in Florida, California, and Hawaii where they are destructive to agricultural plants, especially fruit trees.

Probably the largest alien pet problem is that of house cats, which are abandoned or released by thoughtless owners. Those that survive the first few weeks after release often move into natural areas where they drastically decrease the populations of nesting and migratory birds, as well as lizards and small rodents. They can also be a source of lethal diseases to wild mammals such as sea otters.

FERAL LIVESTOCK Some of the most devastating introductions come from domestic livestock that have become feral or wild and have established reproducing populations. They include goats, pigs, sheep, horses, and burros.

Even though feral livestock are often introduced when they escape through neglect, they are also deliberately released and are often able to exist in a wide variety of habitats and climates. In Hawaii and California, feral pigs have become an extremely destructive force; they root through soils with their snouts, behaving like plows, and uproot native vegetation, an activity which facilitates the colonization of introduced plants and the erosion of hillsides.

ECOSYSTEM MANIPULATION Some deliberate introductions were made by fish and wildlife management agencies when they wished to increase the numbers of a desirable species by changing the food base. Thus, a common practice in reservoirs is to introduce a plankton-eating fish, such as threadfin shad (*Dorosoma petenense*), as prey for introduced predators, such as trout and bass. This is often quite successful, although the shad may suppress zooplankton populations to a point where the juvenile bass, which also require zooplankton, have a hard time surviving. In a natural lake (Flathead Lake, Montana), however, the introduction of a small shrimp as a food organism led to the collapse of the runs of kokanee salmon (also an alien species) up local streams. Kokanee were a major source of food for bald eagles, and when this food supply disappeared, the eagles started foraging more on road-killed animals. As a result, they became road-kill victims themselves. Like biological control, ecosystem manipulation is a risky venture because of the unpredictability in how organisms will respond to changes in their environments.

BYPRODUCT INTRODUCTIONS

Increasingly, species introductions occur as a byproduct of other human activities, especially commerce. Such introductions are often considered to be "accidental" or "unintentional," yet we know they are occurring, and we know they can be prevented. Thus, in reality, they are a form of intentional introduction.

HITCHHIKING Hitchhiking is an extremely common source of alien species. Many of our major pests, from Mediterranean fruit flies to yellow-star thistle to black rats, were carried to North America hiding in ships and airplanes. Airplanes and cargo containers on ships are still major sources of pest introductions, especially insects. At present, thousands of species of small fish and invertebrates are being introduced all over the world through the ballast water of ships. *Ballast water* is water that is pumped into the tanks of cargo ships to help a ship maintain its balance, stability, and fuel efficiency. Modern cargo ships often carry millions of gallons across the ocean and release this water when they get to port. Planktonic plants and animals, including the larvae of fish and mollusks, are pumped in with the water and released as the water is dumped, often thousands of miles away. Clams, fish, shrimp, and zooplankton brought in with ballast water are now causing major problems in the Great Lakes, in San Francisco Bay, and in estuaries worldwide. In the Great Lakes, the tiny zebra mussel has become so abundant that it is clogging water intakes to towns and power plants, shutting some down completely. Keeping the mussels under control is now a major expense. The zebra mussel also competes with native clams, helping to push them to endangered status. Other examples of byproduct introductions include the following:

- The brown tree snake, which has eliminated most native birds from Guam. It arrived in a cargo ship, probably during World War II.

- The tiger mosquito from Asia, a major human disease vector, which came to the United States in water-filled used tires.

- The coqui frog, which arrived in Hawaii riding on ornamental plants. It is the only frog in Hawaii and is now present in the millions. Because of the frogs' incredibly loud calls, especially when they are singing together, coqui are regarded as a major nuisance in urban areas.

ATTRIBUTES OF INTRODUCED SPECIES

Are there attributes of introduced species that allow them to invade successfully and maintain their populations? This is a question that has been asked by ecologists since the 1950s. One attribute that any alien species needs is physiological tolerance for the new environment. Once an alien species is released, in order to successfully establish, it must be able to withstand the range of physical conditions (such as temperature and moisture) in the new area. For example, a pet alligator could not survive in the Arctic because of inappropriate temperatures and other conditions. Therefore, an attribute of many successfully established species is their ability to withstand a wide range of physical conditions or at least the conditions present at their point of introduction.

The ability to subsist on a wide variety of foods is another important attribute of many introduced animal species. This ability allows the introduced animals to find sustenance in areas that have different types and arrangements of food than those found in their native ranges. Successful introduced species also often have high dispersal rates, meaning that they can reproduce and spread their offspring rapidly. The ability to be good competitors against similar species is also thought to be an important attribute.

However, the most important attribute of any alien invader is to be able to live in close association with humans (see next section). If a species can survive in urban, agricultural, or other areas impacted by humans, then it has a good chance of being successful. Once established in altered areas, an alien invader can frequently move into more natural areas. Among the common mammals in most urban parks are black rats, house mice, house sparrows, rock doves (pigeons), and feral cats, all of which are invasive and all of which have amazing abilities to live in contact with humans. Given how rapidly the speed of global trade and transportation is increasing and given the presence of humans in every habitat on the globe, it seems that two general rules apply to introduced species:

1. Any species can be successfully introduced under the right circumstances.

2. Any ecosystem can be invaded by alien species.

ARE SOME AREAS MORE VULNERABLE TO INTRODUCTIONS?

Ecological systems are so complex that it is extremely difficult to predict just where an introduced species is likely to succeed. Charles Elton's landmark book on plant and animal invasions in 1958 concluded that invaders are more likely to establish themselves in areas that have been altered by humans and in areas with relatively simple communities, such as on islands. It is now well established that human-disturbed habitats are more often invaded than undisturbed areas. Disturbed habitats are more easily invaded because the disturbance has already reduced or disrupted the native populations. For example, alien fishes in the western United States can readily establish themselves in new reservoirs that fill after dams are built, but they often have a hard time invading the streams above the reservoirs. It turns out that the streams are generally dominated by native fishes that are unable to survive in the reservoir. Such streams are said to have high *environmental resistance,* meaning that the environment may actually help prevent the invasion.

If disturbed habitats are the easiest to invade, then isolated habitats, such as islands and desert springs, generally suffer the most damage from invaders. Such areas often have simple communities because few species have been able to find their way to them or can survive in such limited areas. The simple native communities of islands are easily invaded because the species in these areas have evolved in the absence of other species and may not have the necessary adaptations to combat competition or predation from novel taxa. Charles Elton referred to this as having a low degree of *biotic resistance.* In desert springs where there are no native fish predators, native pupfishes often have bold, showy, and aggressive courtship displays. As a consequence, individual fish are easily caught

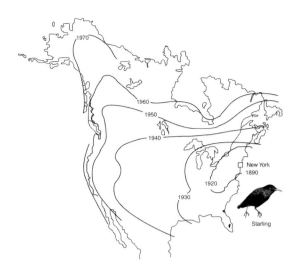

FIGURE 11.3. Map showing the spread of the European starling *(Sturnus vulgaris)* across North America. The original site of introduction was in the Shakespearian Gardens in Central Park, in New York City. (Reprinted with permission from Lockwood, J., M. F. Hoopes, and M. P. Marchetti. 2006. Invasion ecology. Blackwell Press. Oxford.)

and devoured when predatory largemouth bass are introduced into the springs.

Nevertheless, in temperate areas, regions of the *highest* biodiversity may also be the most invaded, because they have diverse and rich habitats and because lots of humans live there as well. In California, the greatest concentration of introduced fishes occurs in the rivers of the Central Valley, where there is also the highest diversity of native fishes. This is also true of plants in Colorado and elsewhere.

CASE STUDIES

We will now present a series of case studies of invasive species, with the goals of highlighting the major themes that have been presented above and demonstrating the diverse nature of invasive species as a conservation issue.

BIRDS, FISH, AND PLANTS IN NORTH AMERICA

STARLINGS AND HOUSE SPARROWS IN NORTH AMERICA

Both the European starling and the house sparrow were purposefully introduced to North America from Europe in the late 1800s. Originating from very small populations, they have both expanded their range to include most of North America. The reasons for the success of these species are not completely known, but one thing is clear: the success of these species has come at a great cost to native birds.

EUROPEAN STARLINGS   European starlings *(Sturnus vulgaris)* are pests in many senses of the word: their populations have expanded dramatically, they have harmed native birds, and they have eaten millions of dollars worth of crops. In 1890 the release of 60 starlings into Central Park in New York City marked the start of what was to become their amazing spread throughout the continent (Figure 11.3). This intentional introduction was perpetrated by an individual who wished to establish in the New World all of the birds that were mentioned in the works of William Shakespeare; the starling is mentioned in *Henry IV*. From the original 60 birds, the population has grown to about 200 million, making it one of the most common bird species in North America. In only a little over 50 years, the bird's populations had spread all the way across the continent; the first starling was officially reported in California in 1942. In the 1960s, many regions initiated eradication programs to control starlings. The programs were successful in killing great numbers of starlings. However, those that survived have been able to build the population back up to enormous numbers, and most control efforts have been abandoned at this time.

Why has the European starling been so successful? Four possible keys may contribute to this species' success, including that it is able to withstand climatic fluctuations, able to compete for food and space with similar species, able to eat a wide variety of foods, and able to coexist with humans. Not surprisingly, the expansion of starling populations has been associated with declines of native birds, especially ones that share similar nesting requirements, such as bluebirds, northern flickers, and woodpeckers. These species nest in tree holes from which starlings actively eject them. In the case of woodpeckers, starlings will wait until they are through drilling a nesting hole and then aggressively prevent them from using it. Starlings are also able to use buildings as nesting sites, so they are able to live in many areas where native species cannot. Like many other introduced organisms, starlings have also been implicated in transmitting diseases, most notably a fungal disease known as histoplasmosis that is harmful to humans.

HOUSE SPARROWS House sparrows (*Passer domesticus*) were introduced to the United States in the late 1800s under the false assumption that they could control insect pests infesting city parks. Unfortunately, someone did not do his or her homework, because house sparrows are largely seed eaters. This sparrow is thought to have become associated with human populations during the rise of agriculture thousands of years ago in Europe and Asia. Thus, it makes sense that their success in North America followed deforestation, the planting of crops, and the general availability of seeds in livestock dung. House sparrows are widely recognized as both aesthetic and agricultural pests and have been implicated in reducing populations of native birds that have similar food and nesting requirements. However, recent research indicates that complete displacement of native birds by house sparrows may be a phenomenon that does not hold up under scrutiny. Regardless, their success provides an interesting case for conservation scientists to study because of a similar species, the tree sparrow, which was also introduced but has not spread across the continent. Differences in size, competitive ability, and genetic variability have been proposed as possible explanations for the disparity in success between the two species.

THE GREAT LAKES AND TWO FISH INVADERS

The Great Lakes system, composed of five major lakes and their connecting channels, is a vast and intensively used body of fresh water. Each of the five major lakes (Superior, Michigan, Huron, Erie, and Ontario) is among the 15 largest lakes in the world, and collectively they contain 20 percent of the world's supply of surface fresh water. The lakes have more than 10,000 miles of shoreline, a total that surpasses the Atlantic coast of the United States by 3,600 miles. The lakes and surrounding land have also been highly altered by human activity during the last several centuries. The aquatic community found within the lakes has undergone major changes, especially within the last century, as alien organisms have found their way into the lakes through human-made canals or been intentionally introduced for recreation.

We will focus on two examples of fish invaders, the sea lamprey and the alewife, both of which have established large populations in the Great Lakes and have had major impacts on native species. The sea lamprey and alewife are both species that originated in the Atlantic Ocean but are able to survive in fresh water as well. This tells us that they would have naturally had access to all of the Great Lakes, if not for the presence of Niagara Falls, which acts as a natural barrier between Lake Erie and Lake Ontario. The falls also at one time prevented merchant ships from moving freely between the Great Lakes and the Atlantic Ocean. Early settlers realized what an enormous economic asset the lakes could be if they could be opened to shipping, and this led to the construction of the Welland Canal between Lake Erie and Lake Ontario in the 1800s. When the canal eventually opened in 1929, species in Lake Ontario, including the alewife and sea lamprey, were provided easy access to the upper four Great

FIGURE 11.4. An adult sea lamprey *(Petromyzon marinus)* attached to an adult lake trout *(Salvelinus namaycush)*. Invasive sea lampreys are major conservation and economic concerns in the Great Lakes. (Photo from the Great Lakes Fish Commission.)

Lakes. Curiously, sea lampreys may not have moved into the upper lakes through the canal but instead were likely established through the use of juvenile lampreys as fishing bait.

SEA LAMPREYS The sea lamprey *(Petromyzon marinus)* is an eel-like fish that preys on large fish (Figure 11.4); it attaches to their sides and extracts blood and other bodily fluids. The sea lamprey was first reported in small numbers in Lake Ontario during the 1830s. The lamprey apparently arrived there in large numbers during the 1870s from the Mohawk-Hudson River drainage in the state of New York, via the Erie Canal, or through its common use as bait. As populations grew into the 1900s, their presence was not felt to a large extent because the native fish populations they preyed on, such as lake trout and Atlantic salmon, had already been severely depleted by commercial fisheries and loss of habitat for reproduction. Once lampreys got around Niagara Falls and through Lake Erie (which was a largely unfavorable habitat for them), they arrived in Lake Huron and Lake Michigan, where they found many prey species. The lamprey attacked native lake trout with a vengeance, and most trout succumbed to the attacks. The abundance of lake trout had already been greatly lowered due to commercial fishery harvests, which made the attack of the lamprey especially devastating. By the mid-1950s, lake trout populations plunged to near extinction in both Lake Michigan and Lake Huron. Lampreys, however, thrived, having lots of alternate species of fish to prey on.

An interesting question is why the sea lamprey was able to have such an impact in the Great Lakes, considering that its effects are minimal in areas where it is native. Native fishes in the Great Lakes were certainly not prepared in an evolutionary sense to withstand attack by lampreys. If these species had evolved with lampreys, then they would have had means to detect and withstand attacks by them. In other inland waters where sea lampreys have been established for thousands of years, such as the Finger Lakes of upstate New York, sea lampreys did not devastate species that they attacked. This is because lampreys are a natural part of these systems, and other species have adapted to their presence.

In 1955 the lake trout was very close to extinction in Lake Michigan and Lake Huron, and the sea lamprey had become such a recognized problem that the governments of the United States and Canada formed a bi-national organization, the Great Lakes Fishery Commission, to develop ways to control or eliminate lamprey. Research was directed toward developing a chemical that harms the juvenile stage of the lamprey that live in streams and rivers. It was discovered that a chemical called TFM (3-trifloromethtl-4-nitrophenol) destroyed many of the larvae and yet had no readily apparent effects on other aquatic organisms except non-predatory lamprey species native to the Great Lakes region. The chemical was widely used to treat lamprey spawning and rearing grounds and had substantial success. The ability to control lampreys has allowed the stocking and reestablishment of lake trout and four other introduced game fish. It has also allowed the recovery of populations of lake whitefish, a favored commercial fish species. Recently, there has been renewed concern over the effects of TFM on other aquatic organisms. The Fishery Commission has had limited success in developing alternative means of controlling sea lampreys and is finding the chemical treatments to be increasingly expensive.

At present, sea lampreys appear to be making a comeback, and there is concern that they will again harm stocks of desirable fish. The lamprey may be developing a resistance to the TFM and are apparently reproducing out in the deltas of streams, which are virtually impossible to chemically treat. One thing is clear: the sea lamprey problem is not going to vanish, and measures for its control will continue to cost millions of dollars a year. Due to human action, the lamprey was able to invade a system containing fishes that were not adapted to the lamprey's style of predation. Interestingly, Pacific salmon species introduced into the Great Lakes have proven resistant to lamprey attacks because these species evolved with the Pacific lamprey, which is similar to the sea lamprey. Chinook, pink, and coho salmon, as well as steelhead, are all now abundant in the Great Lakes, both because they can survive lamprey attacks and because they prey on another alien species introduced into the lakes, the alewife.

ALEWIVES The invasion of alewives *(Alosa pseudoharengeus)* into the upper Great Lakes from the Atlantic Ocean was also remarkably successful. Alewives are small herring-like fish that also gained access to the upper four lakes through the canal system. Unfortunately, their success came at the expense of native fish species such as yellow perch and ciscoes, both of which supported commercial fisheries. Alewives became enormously abundant in the mid-1960s in Lake Michigan and Lake Huron; in fact, they accounted for the vast majority of biomass in the lakes. At that time, this became a problem due to summertime alewife die-offs. Dead alewives literally covered beaches and plugged municipal intakes, often cutting off water supplies to cities and industries. In several extreme cases, the die-offs were so severe that they created public health problems for shoreline communities as well as creating repulsive olfactory problems for miles along the shore. The reasons for the die-offs are not completely certain, but a leading theory is that the alewife, having not evolved in the Great Lakes system, was vulnerable to temperature fluctuations in the lakes. Although millions of alewives died in these events, millions more survived and continued to be a problem. Alewives, then, created major economic costs

by lowering stocks of valuable fish and by necessitating clean-up programs.

Two of the reasons that alewives were able to establish such large populations are their ability to outcompete native fish species for zooplankton and their lack of predation. The sea lamprey and commercial fishing had greatly reduced the numbers of top predators in the lakes. In 1966, coho salmon from Oregon were planted in Lake Michigan and Lake Erie. The following year, Chinook salmon were introduced. The intent of these plantings was to provide fish that would eat the alewife and, as a possible side benefit, to develop a fishery for the salmon. The salmon thrived and grew rapidly to an extent beyond the wildest dreams of fishery managers. The salmon were successful presumably due to their ability to feed on the alewife. For the first time in the Great Lakes, the alewife was a benefit: it was providing forage for salmon and trout that were becoming extremely valuable to the Great Lakes states as sport fish. Three more alien game fish, brown trout, rainbow trout, and pink salmon, were also introduced on the heels of this success.

The alewife population has been declining since the mid-1970s and currently is quite low. This has caused concern because the disappearance of the alewife may mean that coho and Chinook salmon will not have any forage fish to consume, and the loss of these two species would hurt the valuable sport fishing industry. The cause for the decline has been debated by scientists and fishery managers, but it seems that alewives have difficulty living under the variable conditions in the Great Lakes. This could be a classic case where an introduced species establishes a population that then undergoes wild population fluctuations because it is not well adapted to its new environment. It is not known whether the native species that the alewife replaced, such as yellow perch, several minnows, and a cisco species (a fish commonly known as the "bloater"), will ever return to former abundances. There are some indications that these natives are increasing in abundance, but it is not known if they can provide a substantial food supply for the new suite of introduced game fish.

The story of these introduced fishes in Lake Michigan and Lake Huron is an ironic one. The sea lamprey came in and virtually wiped out the only native trout species, the lake trout. This then paved the way for the alewife to establish itself because there were no large fish to control its population, and it was able to efficiently use the available resources. In the 1960s, the lamprey was controlled by chemical treatments, and the alewife population became enormous. To control the alewife, four exotic predatory species were introduced. These alien predators were very successful, and an extremely valuable sport fishery developed for them. The once nuisance alewife had now become highly valuable because it was supporting large populations of game species. However, predation from the salmon and other fish was so intense that the alewife population declined, resulting in decreases in the fisheries. However, fewer alewives meant more food for other fishes, so populations of now rare native species increased.

The entire Great Lakes ecosystem is currently highly unstable. This is due to many factors, and clearly one of the leading ones is that many of the species are not native to the lakes. This case study is a lesson of what can happen when human activities create a system dominated by alien species. Fish populations in the Great Lakes will continue to require intense management for the indefinite future.

TAMARISK

Tamarisk *(Tamarisk sp.)*, also known as salt cedar, is an incredibly invasive, evergreen, dry-loving plant native to the desert riparian (i.e., streamside) areas of Eurasia and Africa. It was introduced into the United States as an ornamental shrub in the early 1800s. Tamarisk is called salt cedar because one of the characteristics of this plant is that it can tolerate very salty soils and has the ability to actually accumulate and concentrate more salts in the soils it lives in. The plant grows as a large bush or small

tree along desert streams and reproduces very rapidly in two different ways. It can asexually bud off new plants from the roots and it also produces hugely abundant tiny seeds (each flower can produce thousands of tiny [<1 mm] seeds), making it typical of *r*-selected species. In addition, tamarisk has very deep roots that can access water from deep underground, making it very resistant to drought. Tamarisk is also resistant to damage from fire and can re-grow very quickly following a burn.

Perhaps tamarisk should be renamed "super invader" because of the plant's incredible ability to disrupt natural ecosystems. One of the main problems with tamarisk is that it alters the structure of native plant communities and degrades native wildlife habitat. It does this through a variety of means, including competing with and replacing native plant species, creating salty soils where other plants can not germinate, monopolizing limited sources of water, and increasing the frequency and intensity of fires. Although individual tamarisk plants do not consume more water than native species, research has shown that large stands of the plant do consume more water than do equal-sized groups of native cottonwoods. Research has also shown that competition occurs when mature stands of tamarisk effectively prevent native species from establishing in the understory, due to shading and elevated soil salinity. The replacement of the native vegetation by this invader also changes the insect and bird communities and may lower a region's overall biodiversity.

Invasion of this plant also strongly correlates with human disturbance to an area, particularly when the disturbance alters a stream's natural flow regime. Typically, streams and rivers have a characteristic pattern regarding how the water flows during the year. For example, in the Mediterranean climate of California, the natural pattern is to have high flows in the winter during the rainy season and low flows during the long, hot summer dry periods. Disruptions to this pattern can occur through human alterations to the watershed such as dams or diversions of water. Human activities that change a stream's flow regime preferentially favor tamarisk and allow the plant to invade and form huge, dense monoculture stands.

Interestingly, the presence of large infestations of tamarisk along riparian areas can actually alter a stream's flow regime further because of the plant's amazing ability to suck up water. It is believed that in large, dense stands, the water-loving roots of this plant next to a desert stream can actually remove enough water so that it can slow or even stop the flow of water in the stream channel! As if that were not enough, large stands also are known to block water movement during floods and can change the way the physical stream mechanics work. To date, tamarisk has taken over large sections of riparian ecosystems in the western United States that were once home to native cottonwoods and willows, and it is projected to spread well beyond the current range.

ISLANDS AND INVASIONS

Although oceanic islands, such as Hawaii, Guam, and Puerto Rico, make up a minute fraction of the Earth's surface area, they are rich in endemic species adapted to their unique conditions. Unfortunately many of these native species have gone extinct in recent years or are threatened with extinction. One of the main factors responsible for these extinctions is the vast number of alien species introduced by humans. In certain cases, the number of introduced species is astonishing. In the Hawaiian Islands for example, it has been estimated that 65 percent of all current plant species have been introduced, along with 20 mammals, 20 reptiles, 20 amphibians, and 50 birds. There are as many introduced insects on the Hawaiian Islands as in the entire contiguous United States! Not surprisingly, a large majority of the documented extinctions of plants and animals within the past 200 years have been native endemic species associated with islands. The role introductions have played in these extinctions can be seen by examining the impact of an introduced snake on Guam, the loss of rock iguanas on Pine Cay in

the West Indies, and the effects of feral animals on the Galapagos Islands.

Understanding why island ecosystems are so vulnerable to species introductions and extinctions requires reexamination of a few evolutionary and ecological ideas we have studied earlier in the text. As we saw in our discussion of island biogeography theory (Chapter 9), because oceanic islands are surrounded by large expanses of water, they are difficult for continental plants and animals to colonize. The few that did make it millions of years ago eventually adapted to their new environment and evolved into an array of unique and wonderful creatures, such as the brilliant honey creepers (birds) of Hawaii. Ecological communities on islands also have fewer species, so their trophic structures and food webs are typically simpler and different from continental communities. They also generally lack top predatory land mammals. This makes island systems relatively easy to invade because they have few predators and little potential for competition. Island species are also highly susceptible to diseases and parasites introduced by humans and their domesticated species because they did not have them in their evolutionary history.

THE BROWN TREE SNAKE ON GUAM

Guam is a small island in the western portion of the Pacific Ocean, about halfway between Japan and New Guinea. Recently, it has been invaded by the brown tree snake (*Boiga irregularis*). The snake is native to Australia, New Guinea, and the Solomon Islands and was apparently introduced via military boats in the late 1940s, perhaps deliberately to control introduced rats. The native forest birds of Guam (a total of 18) have been declining steadily, and several have become extinct. There are many possible factors contributing to the decline, but the brown tree snake is definitely implicated in these extinctions by preying on bird eggs. Thus, as the range of the snake in Guam has expanded, the ranges of native birds have contracted.

What accounts for the success of the brown tree snake on Guam? First of all, there are no significant predators or competitors to limit its population. This situation is similar to what we saw with the alewife and sea lamprey in the Great Lakes. Another reason is that, because of the forest structure and degree of human development, there were few refuges for the birds from the snake. Like many other successful introduced species, the snake has generalized food habits that allow it to maintain high population numbers. Lastly, the snake can go for long periods without eating, a fact that allows it to maintain populations in the face of a low abundance of prey items. The future of the bird fauna clearly rests on whether the brown tree snake can be controlled. However, it will be very difficult and costly to control or eradicate these snakes, so it will likely be years before birds will be able to survive on the island.

FIGURE 11.5. Feral goats *(Capra aegagrus hircus)* are domesticated goats that have gone wild. When feral goats reach large populations, they can pose a significant threat to natural ecosystems. (Photo from morguefile.com.)

ROCK IGUANAS ON PINE CAY

Rock iguanas are a group of related reptile species endemic to the West Indies. Their populations have been steadily declining since the arrival of humans and their associates: dogs, cats, pigs, and mongooses. Several rock iguana species are now extinct, and others are approaching extinction. Pine Cay, a small island in the West Indies, provided a unique opportunity to study the loss of a population of one species of iguana *(Cyclura caurinatus)* in a relatively natural setting following the construction of a hotel and tourist facility in 1973 and the subsequent release of cats and dogs from the facility. The iguana population was reported to have originally been near 15,000 individuals and declined to the verge of extinction just three years after the hotel development. Apparently, feral cats lived on the island in low numbers prior to hotel construction, but they were not a threat to iguanas because they foraged on a large rat population. The new population of cats brought in with the development wiped out the rat population, so the cats switched to eating iguanas. Dogs may have been a more serious threat because they were often observed chasing and catching iguanas for sport. Clearly rock iguanas had little or no defense to the attacks of either cats or dogs. They are now gone from Pine Cay but do exist in protected habitat on some nearby islands (where their populations are unfortunately vulnerable to extirpation by hurricanes). This is yet another example of the loss of a population that did not have the necessary evolutionary adaptations to survive the influence of introduced predators.

INVASIVE SPECIES ON THE GALAPAGOS ISLANDS

The Galapagos Islands are located in the Pacific Ocean about 1,600 kilometers west of the South American continent and just below the equator. The islands have been of interest to biologists because of their unique flora and fauna. The most famous biologist to study on these islands was Charles Darwin. The unique species and communities that Darwin and many others studied are now threatened by large numbers of alien plants and animals. Humans have been introducing species to the islands since the early 1800s, and endemic members of several groups of plants have been reduced drastically and are in danger of extinction. Feral mammals make up a large portion of the introduced animals and have been the most destructive.

Goats on the Galapagos Islands are the most abundant and destructive of the feral animals (Figure 11.5). The first goats came in 1813 and soon spread throughout many of the islands. The speed at which they populated some of the islands is truly remarkable. For instance, on the island of Pinta, a population of several goats increased to about 20,000 in only 15 years. One reason for the goat's success is its ability to thrive at many different elevations on the islands and eat many different plants. Goats also have high rates of reproduction. The impact of goats on the native vegetation has been dramatic. They have managed to remove a wide variety of plants including trees, ferns, shrubs, and herbs. An area on the island of Santa Fe was described as "so over-run by goats that grass and herbs were eaten to the roots during the dry season, exposing the soil to erosion. Bushes were torn up and even the tree-cacti were attacked." Plants on the goat-infested islands are able to grow only in places inaccessible to goats. Several scientific studies documenting the impacts of goats convinced authorities to organize hunting efforts, which began in the early 1960s. Great numbers of goats have been killed, and in areas where they have been significantly removed, many plant species (except those driven to extinction) have begun to return.

Cattle, pigs, and donkeys have also been harmful to native vegetation, although their impact has been less than that of goats. Each of these species has had its own type of impact, depending on the plant species and island considered. Efforts such as fencing, hunting, and poisoning have begun to result in the control of populations of cattle and pigs, but there have been no efforts yet to control donkeys. Cattle have aided in the dispersal of an exotic plant, known as "guava." They eat the fruits of guava and excrete the seeds in different locations. Unfortunately, guava outcompetes native plants. There are also dozens of other exotic plants that have displaced native species.

The native organisms of the Galapagos Islands have been under siege from alien plants and animals for nearly two centuries. It has only been recently that their declining numbers have been appreciated enough to encourage efforts for their protection. It is important for the scientific community to further study the impacts of alien species there in order to help authorities to find ways to save the unique fauna and flora of the Galapagos.

SAN FRANCISCO ESTUARY

San Francisco Bay and its estuary are arguably the most invaded aquatic system in the world, although the eastern Mediterranean Sea is vying for the record. Nearly 300 species of non-native invertebrates and fish have become established there, and new species are becoming established at the rate of one every 12 weeks. In San Francisco Bay, a vast majority of the invertebrates, from clams to zooplankton, are from some place else. Native species are rare. In the upper estuary, a majority of the fish are non-native, ranging from striped bass (from the eastern United States) to shimofuri goby (from Japan) to carp (from Europe). Periodically, a new invader causes dramatic changes to the ecosystem. For example, in 1989 a small new clam from Asia, the overbite clam (*Potamocorbula amurensis*), was found. It quickly took over the bottom of the brackish parts of the estuary, with clams growing on top of clams, creating densities in excess of 10,000 per square meter in places. This clam is now filtering the entire water column in Suisun Bay several times a day, removing most of the planktonic algae and animals from the water column. One result is a drastically reduced food supply for fishes, especially the juvenile stages that use the bay as a nursery area. Another problem is that, because they filter the water so effectively, the clams are concentrating the heavy metal selenium in their flesh. This may ultimately create a toxic problem for animals that feed on the clams, such as sturgeon and diving ducks.

The major source of recent introductions, such as the Asian clam, has been ballast water, because San Francisco Bay is a major port. Millions of gallons of water containing millions of organisms are dumped into the bay every week. They continue to be introduced because the shipping industry does not have to pay for the damages caused by introduced species such as the clam (or the zebra mussel in the Great Lakes). There are solutions to the problem but they involve changes to ship operations as well as continuous monitoring of the ballast water of incoming ships. In general, legislation to regulate ballast water dumping has failed to solve the problem. Stricter legislation has been passed, but it remains to be seen how well it will work.

CONCLUSIONS

Alien species have been introduced for many different reasons and have had multiple impacts on the places they invaded. While many alien species appear to be benign from the perspective of their effect on native species, rarely are studies strong enough to be able to demonstrate a lack of impact. In any case, we typically measure the level of impact based on how much a species hurts our economy. For instance, zebra mussels, sea lampreys, and starlings are considered to be major problems because they affect human water supply, fisheries, and crops, respectively, resulting in millions of dollars being spent for their control. These same species also have large-scale negative effects on native animals. In addition, introduced species sometimes cause harm that goes unnoticed because there is no direct economic threat, or strangely, introduced species may sometimes even provide some amount of economic benefit (despite the ecological damage).

Increasingly, we are living in world where many of our ecosystems have alien species as a significant component of their overall biodiversity. Interestingly, many of these invaders tend to belong to a relatively small number of common species. This has the effect of producing what is called *biotic homogenization,* which is the process of making different ecosystems more similar to each other through the widespread introduction of the same set of species. There are two major effects of this trend. First, the aliens may make natural systems much harder to manage because they can significantly alter ecological interactions, especially if new species are moving in on a regular basis. Second, worldwide biotic homogenization is making the planet's diverse ecosystems

less interesting, as natural systems lose their distinctiveness.

Introducing species can enhance local diversity in the short term, but the problems that alien species can create (competition, diseases, unstable populations, etc.) threaten diversity over the long term and from a global perspective. Peter Moyle and Hiram Li, in a paper discussing the impacts of introduced fishes, put the issue in perspective through the use of a parable:

> There is a long and honorable tradition in western culture, dating back at least to the Romans, of tinkering with fish faunas by adding new species. This tinkering is part of a much broader tradition of tinkering with nature, to "improve" on it. The moral and mechanical problems that are encountered when trying to improve on nature were dramatically illustrated in Mary Shelley's famous novel (published in 1818), "Frankenstein, or the Modern Prometheus." In this story, Count Frankenstein, a dedicated scientist, attempted to create an improved human being but soon discovered, to his mortal distress, that he had created more problems than he had solved. Most of his problems stemmed from focusing on the solution to a narrowly perceived problem without considering how the solution (the monster) would fit into society at large.

This parable of the Frankenstein effect suggests that we should be concerned about long-term and broad-scale consequences of our actions. Although society perceives many alien organisms as beneficial, introductions are collectively harmful to biodiversity, to the planet, and ultimately to society as a whole.

FURTHER READING

Bright, C. 1998. Life out of bounds: bioinvasion in a borderless world. W.W. Norton and Co. New York. *A "popular" introduction to the problem of invasive species.*

Cox, G. W. 1999. Alien species in North America and Hawaii: impacts on natural ecosystems. Island Press. Washington, DC. *As above, but focused on North America.*

Crosby, A.W. 1986. Ecological imperialism: the biological expansion of Europe, 900–1900.

Cambridge University Press. Cambridge. *A fascinating look at how alien species got moved around by our culture.*

Elton, C. S. 1958. The ecology of invasions by animals and plants. University of Chicago Press. Chicago. *The first important book on the subject, and still highly readable.*

Lockwood, J., M. F. Hoopes, and M. P. Marchetti. 2006. Invasion ecology. Blackwell Press. Oxford.

This textbook is designed for upper-division biology majors, but much of it is understandable to non-scientists, and it is a good starting point for people interested in this subject.

Todd, K. 2001. Tinkering with Eden: a natural history of exotics in America. W. W. Norton and Co. New York. *Written for a general audience, this book is a cautionary tale that is both illuminating and entertaining. Todd is a terrific storyteller.*

12

Restoration Ecology

So far in our exploration of the science of conservation, we have been dealing a lot with species loss and extinction, which are fairly negative and sometimes disheartening topics. Although the field of conservation biology exists in good part because of the extinction crisis, a growing aspect of conservation science is focused on finding solutions, especially within the emerging field of *restoration ecology*. In this chapter, we explore what restoration ecology is and what it is not; we also examine a number of examples of both large- and small-scale restoration projects to give us a flavor of where this new science is headed. Many people are pinning their hopes to this field, believing that this discipline carries the opportunity to fix some of the problems we have created. As we will see, restoration may allow us to partially return native plants and animals back into our urban landscapes, but it may not be the holistic environmental fix we might hope for.

RESTORATION DEFINED

Ecological restoration is often thought of as *the process of intentionally altering a location in order to reestablish an historic ecosystem, including the original ecological structure, function, and diversity of the place.* There are a few things to notice about this definition:

- Restoration is considered a process, not an event. As we will see, restoration takes time and is not done in one fell swoop.

- Restoration is an ecological process involving a location's ecological structure and biodiversity and therefore requires significant environmental knowledge of an area prior to beginning.

- Restoration directly involves human beings in the process, which is different from some forms of conservation we have seen previously (i.e., the establishment of reserves and protected areas).

- Restoration sets out to reestablish historic ecosystems, which is a noble goal that is unfortunately impossible in a practical sense, given the pervasive impact of humanity on the world.

A fundamental philosophical problem with the process of restoration is that temporal change and environmental variation, both in the short and long term, have always been a part of the way natural ecosystems function. In addition, human-mediated change has also been part of most ecosystems for thousands of years. In fact, with the pervasiveness of forces such as global climate change, alien species, and the expanding human population, restoring an area back to its original ecological system is not realistically possible or even desirable. Thus, the definition of restoration given above mistakenly assumes that we can return a location back to a mythical steady-state environment prior to human beings.

Take, for example, the Sierra Nevada mountain range, which forms the backbone of California. In 1997, the Sierra Nevada Ecosystem Project attempted to make recommendations for protecting the mountains from too much human use and for finding ways to restore key parts of the ecosystem, such as large tracts of forest with big trees. Almost immediately, one of the problems facing project organizers became trying to determine how far into the past the restoration project should aim for. In other words, what should the restoration target be, and how far back in history should we look? Should the goal be to restore the current forests to the environment that existed before the first Native Americans arrived in North America, or should the restoration goal be to restore the conditions that existed prior to the Gold Rush? The problem with the latter scenario is that the various Native American peoples, who lived in the Sierra before 1850, intensely managed the forests with fire and other tools to favor vegetation that produced food and fiber for them. So when European settlers arrived on the heels of the gold rush, the forest they encountered and thought "pristine" was in fact an artifact of the native people's management practices. Forest management by Native Americans produced a complex mosaic of habitat, including relatively open forests with an abundance of large old trees. Thus, restoring the forests to this previous era's management regime might be neither desirable nor possible.

Another problem with restoring the Sierra Nevada to a long-past era is that no one today really wants large populations of giant grizzly bears roaming the mountains outside of Fresno, Tahoe, or Reno, as the bears once did. Too many people currently live in the mountains for this to be a realistic goal. So instead, for the Sierra, the default forest restoration goal is one of a spatially open, fire-prone forest with big trees but without grizzly bears; this is very different from the idealistic and lofty goal identified in the definition above.

Thus, a more realistic definition for ecosystem restoration would be *the process by which humans bring a human-degraded ecosystem to a more desirable state, in order to protect ecosystem services, native species, aesthetic qualities, and the potential for long-term sustainability with minimal human input.* This definition reflects the fact that, no matter what happens, human beings are the main drivers of ecosystem change; thus, we must take responsibility for perpetuating and maintaining remnants of historic ecosystems worldwide. Realistically, these restored systems must also provide significant ecological and ecosystem benefits to humans in order to have any hope of persistence over the long run. Thus, in the case of the Sierra Nevada forests, recreating open forests with big trees provides the following benefits:

- An ecological system with a great deal of fire protection for human infrastructure.

- An improvement in the ability of the forest to produce a stable water supply.

- The potential for a steady supply of timber to be harvested sustainably.

- Critical habitat for rare creatures such as flying squirrels, spotted owls, wolverines, and pine martins.

- A habitat that we humans enjoy being in, for living or for recreation.

Yet the restored Sierra forest will also contain numerous alien species, such as European grasses under the trees and non-native trout in the streams. One way of approaching this type of integrated restoration is to manage the system *with* change (rather than against it), while fully integrating humans into the restored eco-system. This philosophy of restoration has been called *resilience thinking* by Brian Walker and David Salt, in their small but provocative 2006 book of that name.

WHAT IS RESTORATION?

Another way to approach a definition of restoration is to ask what it is that restoration ecologists do. From our definitions above, we can see that they attempt to reestablish the natural components of a site, which often involves focusing first on the vegetation. Why focus on the vegetation? It appears that ecosystems may work somewhat like the maxim from the movie *Field of Dreams:* "If you build it, they will come." Building it, in this case, means putting a community of primary producers in place and helping them become established. Secondly, restoration ecologists want to create systems that are persistent through time, although how much they are allowed to change will depend on the degree of human intervention. Eventually, it is desirable for the restored systems to work and function on their own with minimal human inputs. That being said, restoration practitioners also recognize that the process relies heavily on human action that generally dominates the early stages of a project. Humans do restoration work and are therefore a vital part of the process.

An interesting way to visualize the restoration process can be seen in Figure 12.1. Here, we have essentially a cartoon-like graph showing the ecological underpinnings of the restoration process. First, notice the two axes. The *x* axis represents the number of species (or the biodiversity) as a measure of ecological complexity, and the *y* axis represents the biomass of living organisms as a measure of ecological function. If we are given a degraded ecosystem (i.e., an abandoned agricultural field), it generally has both low biomass (not much growing on it) and a low number of species present (typically a few species of weeds). The goal for the restoration process is to move the degraded system toward the restoration target in the upper right of the

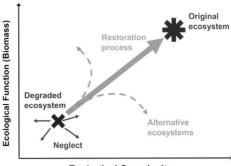

FIGURE 12.1. Graph depicting the hypothetical process of restoration using two ecological measurements: complexity (i.e., number of species) and function (i.e., biomass). Degraded ecosystems tend to have both a low number and a low biomass of species. The restoration process is one of intentionally moving the system toward more ecological complexity and function. Note that the restoration trajectory is not the only possibility for the degraded system, because a system usually cannot be restored completely to its original state. Alternative states result when alien species are integrated into a system and native species are extirpated, or when irreversible changes have occurred to the landscape. (Modified after figure from Bradshaw, A. D. 1987. The reclamation of derelict land and the ecology of ecosystems. In Restoration ecology: a synthetic approach to ecological research. W. R. Jordan III, M. E. Gilpin, and J. D. Aber, eds. Cambridge University Press. Cambridge.)

figure. This target is generally richer in species and contains more biomass. The restoration process is represented by the arrow connecting the two areas. Notice also that there are other potential paths that the degraded ecosystem can follow. If left on its own, without removing the factors degrading the system (e.g., pollutants, intense land use, etc.), the system will either gain or lose species but will generally not recover and progress to the original system's conditions. However, if the major stressors are removed, sometimes a surprising degree of recovery is possible. For example, many dense forests tracts in the eastern United States were once pasture-land that was abandoned 150–200 years ago. It is also possible to move the degraded ecosystem on entirely new trajectories represented by the dotted arrows in Figure 12.1. This usually happens when a few species of plants, typically alien species, take over a region in the absence of normal controls, resulting in an entirely new system with more biomass but low diversity or vice versa. The closer the restoration scientists can follow the trajectory toward the target, the more quickly the desirable restoration conditions will appear. So in this scheme, restoration is a process that moves the degraded system toward the restoration target.

THE PAST INFORMS THE PRESENT

Philosophically, as we saw above, restoring the ecosystem function of an area raises a host of challenging questions. Partly this happens because the process depends on knowledge and understanding of what was present in the past and how it functioned, two things that sound simple but are incredibly hard to know precisely. For example, if we wanted to restore the landscape that once held the forests of the Iberian Peninsula (modern Spain and Portugal), how could we know what they looked like prior to human occupation 40,000 years ago? There were no scientists on site to record the ecology of the area, to document food webs, or even to make a list of the species that were present. We can find some clues from fossil and pollen records, but we can never know for certain that we are able to identify even half of the species involved in that area's prehistoric forest ecology. In addition, if we can not even get a complete species inventory (i.e., a measure of biodiversity or species richness) for a current extant ecosystem, then how would it be possible for scientists to restore an ecosystem's complex functioning? We have enormous difficulty describing and understanding modern ecosystems that we can study and experiment with, let alone ecosystems of the past. Remember the complicated food web involving the Atlantic cod that we saw in Chapter 6. To even assemble the food web picture took decades, and scientists still do not completely understand how all the players in that food web interact with each other and with their environment.

As we mentioned above, another important consideration for restoration projects is to decide how far in the past to make our restoration target. Should we set our restoration sights on 50 years ago, before the human population got so large, or perhaps 200 or 2,000 years ago, or before humans were even on the scene? What state are we trying to restore a system to? In other words, how do we deal with the fact that we know ecosystems are dynamic and change through time? Should we even bother to restore something that is constantly in a state of flux? These questions are difficult and fundamental to this burgeoning science and are as relevant to the people doing the work on the ground as they are to the planners and funders of the restoration projects. For now, it is enough for us to realize that restoration may look simple on the surface but may be extremely challenging for us to achieve.

A NEW BRANCH OF SCIENCE

When we say that restoration ecology is a new branch of science, we mean that, although people may have been doing some forms of restoration for a long time, the scientific community has only recently brought together the tools, practices, and language of restoration under one roof and called it a new field by adding ecology to the name. You might wonder, When does a science become a new field, and how new is new? We can address both of these questions by looking at one aspect of actually doing science, namely the scientific literature.

When scientists conduct research, they often find new, interesting, and potentially useful information they want to share with each other and the world at large. But before they can share the information, they need to make sure that their work is repeatable and free from factual and judgmental errors, so they send descriptions of their work to scientific journals. Prior to publishing the results, the journals send the manuscripts out to other professional scientists, who are also experts in the field. These "peer reviewers" read and comment on the research. If the reviewers find mistakes or disagree with the methods or results, then the research is either not published or is sent back

TABLE 12.1

Conservation Biology Compared with Restoration Ecology

TRAIT	CONSERVATION BIOLOGY	RESTORATION ECOLOGY
General outlook	Threat of permanent biodiversity loss	Long-term recovery of ecosystems
Research focus	Genes, populations, species	Communities, ecosystems
Taxa of interest	Animals (vertebrates)	Plants
Scientific tools	Population viability analysis, mathematical models	Succession, field experiments

to the authors to correct their mistakes. This process, although time consuming and sometimes quite frustrating, ensures that work published in scientific journals contains results that can be relied on. This also means that when you read something in a scientific journal, you can have confidence that it represents some version of the truth, which is not true for other forms of literature.

Restoration ecology, as part of conservation science, became a distinct discipline during the late 1980s as practitioners started using more systematic and repeatable methods and began to test their ideas as rigorously as possible. They joined together and formed new scientific societies and began publishing journals dedicated solely to the science of restoration. The current trend in restoration research is an increasing focus on long-term recovery of communities and ecosystems, while treating the large-scale restoration projects as individual experiments. Yet, if you think about this, if the restoration projects are experiments, then it is possible for them to fail, and in many cases, some amount of failure is even expected, because failed experiments are something we can learn from. Today's restoration projects that are not 100 percent successful are therefore paving the way for tomorrow's restoration successes.

CHANGE IN OUTLOOK

Some people feel that restoration ecology is properly thought of as a subdiscipline of conservation science, whereas others feel that it is a field in its own right. Although this distinction is not of vital importance, it is a useful exercise to examine the similarities and differences between the two areas. Truman Young at the University of California, Davis, published a paper in the journal *Conservation Biology*, where he directly compared the two fields. His central idea was that the two disciplines represent fundamentally different worldviews and, as a result, have striking differences in terms of their general attitudes. The gist of his argument can be summarized in a table comparing the two fields (Table 12.1). According to Young, the philosophical outlook of conservation biology is one of biodiversity loss and extinction, whereas restoration ecology suggests the possibility of long-term recovery. Whereas conservation biology has focused much of its research energy on the lower levels of ecological organization such as genes, populations, and species, restoration is often directed toward the larger levels of communities and ecosystems. Within conservation science, there has been a distinct focus on preservation of animal taxa, particularly vertebrates, whereas restoration work is often focused more on plants. Conservation biologists have utilized tools such as mathematical models and population viability analyses (PVA), whereas restoration ecologists use tools such as field experiments and ecological succession.

For some people, the picture that emerges from this comparison is that restoration ecology paints a more hopeful and optimistic outlook on the state of the world. If restoration ecology is successful to some degree, then some of the doom associated with conservation work may be reversed. In fact, at some point in the near future, the vast majority of intact natural eco-systems around the planet will either be under some sort of protection (i.e., as parks, reserves, easements, etc.) or will have been severely altered by the forces of development. At that point in time, the science of conservation, with its focus on how to protect these systems, will be forced to shift its approach. What will be left for scientists to work with will be degraded land-scapes in need of ecological restoration. Our goal at that point will be to integrate the natural world into the constructed world humanity has built (see the reconciliation ecology section at the end of this chapter). In the meantime, we should protect as much of the remaining natu-ral systems as possible and continue to develop the tools we need for large-scale restoration and integration projects.

CONSIDERATIONS FOR RESTORATION

When a restoration project is conceived, there are a number of important issues to consider before the work can go forward. For example, a major factor in restoration work is the cost of completing the project. Are the funds available to see the entire thing through to completion? This can vary by the type of project. Current work in mine restoration often has considerable financial backing because, in order to get legal permission to mine an area, the mining com-pany often has to put up a significant amount of bond money. This money is used at the end of the project to restore the degraded mining site and the surrounding areas. As a result, some of the largest and best-funded restoration projects are for abandoned mine sites, although this is a fairly recent development stemming from hor-rible toxic waste and other environmental prob-lems created by many untreated mine sites. Unfortunately, the reality is that most restora-tion projects are woefully underfunded.

A second consideration is the speed at which the restoration can proceed. The length of time it will take to restore an area depends on the assemblage of organisms we want to restore. For example, a grassland restoration

project may be completed in a relatively short time because grasses tend to grow very quickly (although the herbivores and grazers make take longer to come back). The major structural components of a grassland restoration project may be put together in just a few years, so the project *looks* like a natural grassland. However, it would be arrogant of us to assume that the native grassland has actually been restored until it has gone through many years with the pro-cesses of natural variability and change being allowed to work. For example, too little burning of a grassland may allow for invasive grass spe-cies to become dominant, or it may cause the loss of a key herbivore (e.g., prairie dogs). The result of this small change (e.g., too little burn-ing) may be a much less complex ecosystem than the natural or unaltered state. "True" res-toration of a grassland is much more likely to succeed if the restoration area is associated with a larger tract of more pristine grassland.

But the time it takes to restore a grassland is likely to seem very short compared to how long it would take to restore an old growth forest in the Pacific Northwest. Old growth is defined by having trees between 150 and 500 years old, so it clearly would take decades, if not centuries, to fully restore such a forest to some state approach-ing its original condition. Or consider the prob-lems of restoring a coral reef that has lost most of its hundreds of species of "old growth" coral and associated fishes. How do you bring all the obvious big species back, as well as all the other, more obscure species that are missing? Thus, it appears that the more biologically complex a system is, the more unlikely it is that we can restore it to anything approaching its original state. As a consequence, if we are not able to gather all the players, it is likely that our restora-tion efforts will be missing some key ecological elements that would allow the restored ecosys-tem to function properly.

As an example, let's turn to an unlikely restoration ally, namely mycorrhizal fungi. Mycorrhizal fungi are detritivorous plant-like organisms that live in and throughout many types of soil and form very intimate associations with some plant roots. These fungi have also been shown to sometimes play an important ecological role in the soil community by helping certain plants acquire nutrients (i.e., they engage in a form of mutualism with the plants; see Chapter 6). When the correct group of natural fungi are present in the soil, the native plants tend to grow better and faster. However, until recently, most restoration projects failed to include mycorrhizal species in the restoration plan. Faster growing and healthier plants provide a better chance for restoration success; therefore, it may be important in some situations to include unlikely organisms such as fungi in a restoration plan.

Finally, another important consideration revolves around issues of persistence, or how long the restored community will last. If a restoration project requires intense human intervention (i.e., mowing, weeding, watering, or burning) then long-term success depends on a dedicated human commitment to maintain the system. On the other hand, if the new community requires minimal maintenance, then it may persist longer into the future, surviving periods of low funding and low community interest.

CASE STUDIES

At this point, it may seem that restoration ecology is a bit ambiguous and poorly defined. Partly, this is a result of the newness of the discipline. Many restoration projects are essentially writing the rules as they go. In addition, the nature of restoration work makes it hard to develop concrete theories. Each restoration project faces different obstacles and challenges, just as each restoration location has its own set of ecological constraints. This does not mean that there will never be a textbook for ecological restoration (in fact, there are a number of them out there). Instead, it points to the need to review restoration case studies and try to extract generalities from them. We will look at four different restoration projects to give you an idea of where restoration ecology is heading: the riparian corridor along the Sacramento River

in California, the Yellowstone ecosystem in the western United States, the Guanacaste National Park in Costa Rica, and the Everglades ecosystem in Florida.

THE SACRAMENTO RIVER

The riparian forests that grow along the banks of a river or stream are intimately tied to the ecology of the aquatic environment. Streams feed the riparian forests, and the forests in turn feed the streams. One of the major environmental factors that drives the ecology of the riparian forests is the flow of water and associated periods of flooding. When water is removed from a riparian system (through canals or diversions), or when natural floods are prevented (by dams or levees), the riparian forest suffers. Unfortunately, many of our large cities and urban areas are situated on stream or river systems, forcing us to protect ourselves from natural floods and redirect the flow of water for human uses such as agriculture. This is true for the Central Valley in California, which was described by Mark Reisner, the author of *Cadillac Desert,* as one of the most transformed landscapes on the planet. As a result, the riparian forests that once stretched over a mile wide on either side of the valley's main rivers, the Sacramento and San Joaquin, are largely gone and have been replaced by huge tracts of agricultural fields, which allow virtually no vegetation other than crops to grow on them.

In the late 1980s, California State Senate Bill 1086 was passed, mandating that the state "maintain and restore riparian habitat of the Sacramento River and the natural processes on which it depends." In conjunction with this, some farmers in the northern part of the Sacramento Valley were willing to sell off portions of their land adjacent to the river because frequent flooding made the land not economically viable to farm. The Nature Conservancy (TNC), a nonprofit conservation organization, began to buy up some of this land in order to plant riparian forests. This effort eventually grew into a gigantic endeavor that is known as the Sacramento River Restoration Project,

FIGURE 12.2. Restoration planting of both overstory and understory riparian plants along the Sacramento River, in California. (Photo by Karen Holl.)

which involves researchers from the California Department of Fish and Game, the U.S. Fish and Wildlife Service, the University of California at Davis, California State University at Chico, TNC, and many others.

The scope of the project is impressive, in that the aim is to restore the riparian forest along a 100 mile section of the river. The project has three goals: (1) to acquire flood-prone land along the river, (2) to restore the riparian forests by planting native vegetation (both overstory and understory plants), and (3) to restore a natural flow regime in the Sacramento River, which allows only a limited amount of natural winter flooding and river meander. In planning this work, project partners want to restore ecosystem health to the riparian and river ecosystems as well as to accommodate human use of the area.

So far, the project has been very successful; it has acquired large amounts of land and has begun the enormous task of planting native vegetation (Figure 12.2). It is likely that this section of the Sacramento Valley will begin to resemble the riparian forests that were present along the river prior to the Gold Rush in the 1850s. Yet at the same time, the forests will also support many non-native plants (such as English ivy) and animals (such as common carp, black rats, and feral cats), which will probably require continuous control efforts. Some key animals, such as grizzly bears and tule elk, will be missing from the forest for the indefinite future.

It is worth noting that considerable economic benefits are likely to be realized from this project, including increased recharge of groundwater, reduced cost of downstream flood protection, improved salmon fisheries, and reduced costs to protect endangered species. At the same time, the aesthetic benefits realized from a riparian forest dominated by large trees and many birds are considerable.

YELLOWSTONE

The grey wolf *(Canis lupis lupis)*, a top mammalian predator, had an historic range that covered most of North America, but by 1900 the species had been hunted to extirpation throughout most of the lower 48 states. When the Endangered Species Act (ESA) was passed in 1973, the grey wolf was listed as threatened, protecting the species from further eradication efforts. The idea of reintroducing wolves to Yellowstone began in 1966, when park biologists noticed a large increase in the population of elk, the wolves' main prey. Over the following decades, scientists began to notice other changes to the greater Yellowstone ecosystem as a result of the loss of wolves as top predators. For example, the willow and aspen forests were being overgrazed and removed by the huge populations of elk. The low numbers of young trees, in turn, had a negative impact on both moose and beaver, which required them for food. As beaver left the area, the loss of their dams on waterways changed how the stream systems functioned and also affected fish populations. In addition, with the removal of wolves as top predators, coyotes stepped into the role, and their numbers increased. Increased coyote predation on smaller animals such as mice and voles had a negative impact on red fox numbers as well as on avian predators such as hawks and eagles. The loss of wolves also meant fewer large animal carcasses left from their pack hunts, which were a significant portion of the diet of grizzly bears and wolverines. Less food for these large predators meant less chance of successfully rearing young, and so populations of these large predators declined. All in all, the loss of wolves had a profound impact and produced a ripple effect (i.e., a trophic cascade; see Chapter 6), which was felt throughout the entire greater Yellowstone ecosystem.

Despite this, the idea of wolf reintroduction was vehemently opposed by the ranching industry because ranchers felt the return of wolves to the area would destroy their livestock. This argument became the biggest hurdle to wolf restoration until an environmental nonprofit group called the Defenders of Wildlife suggested a "wolf compensation fund" in 1987. The idea was that the nonprofit group would raise money through donations to reimburse ranchers the full market value for any livestock lost through wolf predation. On the heels of this proposal, a recovery plan under the Endangered Species Act was eventually finalized, and on January 5, 1995, wolves were restored to the Yellowstone ecosystem for the first time in almost a hundred years.

The restoration has been a resounding success; as of 2008, there were an estimated 124 wolves (12 packs) in the park, which is well beyond the target goal of 30 breeding pairs. There has also been minimal loss of livestock with the return of wolves, and the Defenders of Wildlife have paid out approximately $500,000 in compensation over 10 years' time. The biggest measure of success for this restoration has come from the increased biodiversity in Yellowstone National Park as a result of the wolf reintroduction. The wolves have put the elk population in check, allowing for more young aspen and willow trees to grow. This, in turn, has allowed beaver to return to the park for the first time in nearly a century. Red fox populations are increasing, as are grizzly and wolverine populations. It's worth noting here that the restoration of the Yellowstone ecosystem did not focus on the lowest trophic levels (i.e., primary producers) but instead worked at restoring a balanced food web from the top. Overall, the restoration of the Yellowstone National Park ecosystem through the reintroduction of a top predator has been very successful.

FIGURE 12.3. Guanacaste National Park is a national park in the northern part of Costa Rica. The park was created to provide an ecological corridor between the tropical dry forest to the north and the rainforests to the south, and it allows many species to migrate seasonally between these areas. (Photo by Elena Berg.)

A bigger challenge now lies in trying restore historic ecological function to the greater Yellowstone ecosystem, including regions outside the park boundaries. Yellowstone National Park itself is not really big enough to support self-sustaining populations of wolves and other predators that require large amounts of territory. Although there are large tracts of public land around the park, they are interspersed with private land, especially ranches and tourist villages, on which there are major conflicts with predators (and other wildlife). Ranchers, in particular, claim that not only do wolves kill their cattle, but their presence and harassment also makes their cows unable to gain much weight. As a result, some states in the region have recently instituted wolf hunts for sport.

GUANACASTE NATIONAL PARK

Guanacaste National Park covers an area of approximately 83,000 acres in the northwestern portion of Costa Rica. This park is different from many national parks, in that the site was not chosen due to the pristine or intact nature of the habitat. In fact, much of the land within the park was heavily grazed and farmed before being restored to native forest (Figure 12.3). This area was chosen for restoration because it is an important corridor for animal movement between the dry tropical forests to the north and the rainforests of the south. The park's existence was heavily lobbied for by Dr. Daniel Janzen of Florida State University because of this important ecological role it plays. Biodiversity surveys in the park have identified over 140 species of mammals, over 300 species of birds, 100 species of amphibians and reptiles, and over 10,000 species of insects.

When the park was established in 1989, 14 farming families sold their farmland to the park but essentially remained living on the land as stewards of the area. Slowly, the farmland reverted back to rainforest, and a more natural ecosystem was restored. It is a very different kind of national park in another way as well, because it is supported by a nationwide environmental program. The idea of the program is to get the local community involved in the park and to feel that it has a stake in the park's success. Providing a strong education program is a large part of this strategy. Nationally, there are programs for schoolchildren, in which students spend whole days in the park learning the science behind ecology and restoration. The park also has internship programs for older students that sometimes lead to professional employment later in life. Local landowners are allowed to graze livestock in certain areas of the park in an environmentally sustainable manner. Scholars who visit the park live in home-stays with local families and therefore also experience a close connection to the management of the park. Essentially, the park is an experiment in both ecosystem restoration and community-based resource management and conservation. The restored ecosystem is very similar to the historic ecosystem that was present but includes a stronger human component than was originally present.

THE FLORIDA EVERGLADES

Everglades National Park is the largest subtropical wilderness area in the United States and the largest wilderness area east of the Mississippi River. The park was created

FIGURE 12.4. Everglades National Park, in Florida, is the largest subtropical wilderness in the United States. Although a majority of the animals and plants are native, alien plants and animals today play a role in this ecosystem's function. (Photo by Mark Rains.)

to protect the Everglades ecosystem, although it contains only about 20 percent of the total system. Currently, there are 15 federally designated threatened and endangered species found within park boundaries, including the Florida panther and the American crocodile. The park also contains a large diversity of habitat, including the largest mangrove system in the western hemisphere (Figure 12.4), and it is an important breeding ground for millions of migratory waterfowl.

Despite these impressive statistics, the entire Everglades ecosystem is under threat from a severe lack of water. In its natural state, the entire southern part of Florida is a very slow moving river, with water flowing in a southward direction and covering approximately 4 million hectares. Since the 1850s, people have been trying to control, divert, and drain this gigantic river system for agriculture and urban uses. With the completion of the of the Central and South Florida Project in the 1950s, there were over 1,000 miles of levees and 750 miles of canals built in the region. An unintended consequence of this development has been to essentially halt the large-scale flow of fresh water southward into the park. As a result, the park and its fragile ecosystem are slowly drying up and are critically threatened by this loss of freshwater flow.

In order to protect this unique system, President Bill Clinton signed the Comprehensive Everglades Restoration Plan (CERP) in 2000 before leaving office. The scope of the plan is enormous, as it covers 16 counties and an 18,000 square mile area of south Florida. The plan will take upwards of 30 years to construct, at a projected cost $7.8 billion. The goal of the plan is to capture some of the fresh water that flows unused to the oceans and redirect it to areas in the Everglades to restore and revive a dying ecosystem. It has been called the largest restoration project in the world, and if it is successful, it may be the first time an entire ecosystem, including its ecological functions, will be restored. Currently, this restoration project is mired in legal and political difficulties, and a fully restored everglades ecosystem is still a long way off.

RECONCILIATION ECOLOGY

The four examples above are clearly much more than simple restoration projects because all involve a large human component to the ecosystem, and the restored ecosystems (even Yellowstone) have significant differences from the historic systems that existed there, including many alien species. Recognizing this complexity, Dr. Michael Rosenzweig, of the University of Arizona, has proposed the development of what he calls *reconciliation ecology,* which is a close relative of restoration ecology. His basic idea is that biodiversity protection will require us to incorporate animals and plants we treasure back into our human-dominated systems. There are many species that can thrive in urban systems if we provide them with the necessary ingredients to complete their life cycles. For example, if we want to include songbirds in our urban landscape, then we need to first plant appropriate trees and bushes for nesting and feeding into areas that the birds can use. This is not difficult to accomplish and can make an enormous difference in terms of the number of bird species found in our cities and towns.

One dramatic example of this can be seen in the return of falcons and hawks to Central Park in downtown Manhattan. Many New Yorkers would not have believed it possible, but a small population of top avian predators has made a home in one of the most densely populated areas of the United States, where they prey on pigeons and other urban birds, showing that some wild species are able to live in contact with human landscapes.

Reconciliation can also work on a large scale. For example, outside the California city of Sacramento is the 24,000 hectare Yolo Bypass, a wide, shallow channel that was designed and constructed to act as flood protection and bypass large flows around Sacramento. Built in the 1920s and 1930s, the bypass has been very successful in protecting the city at a relatively low cost. Although historically, the main use of the bypass has been for agriculture, recent studies have demonstrated that it is also very important for fish and wildlife species. When it floods with the winter rains, huge flocks of ducks and geese forage in its productive waters. Juvenile salmon that enter the flooded areas to feed have been shown to grow faster and survive better than their compatriots that stay in the main river. Sacramento splittail, a large native minnow that requires floodplains for spawning, depends on the Yolo Bypass for spawning and rearing of its young due to the loss of its natural spawning grounds.

Recognition of the biodiversity value of the bypass has resulted in the ecological restoration of sections of the bypass as wildlife refuges, whereas other parts are maintained for productive agriculture. Interestingly, the Yolo Causeway, a large highway bridge that crosses the bypass, supports the largest colony of bats in California. These bats have found a location to roost in an area swarming with insects for them to eat that is almost in the direct center of one of the largest and most productive agricultural landscapes on the planet.

CONCLUSION

By itself, restoration ecology is not ever going to be enough to solve the planetary biodiversity crisis we are facing, but restoring damaged communities can go a long way toward protecting biodiversity. Yet if we truly want to protect and restore biodiversity, we will eventually need to look beyond pristine landscapes and find value in places such as abandoned agricultural fields in Central America, degraded and abused riverine systems like the Everglades, and even artificially constructed landscapes such as the Yolo Bypass. We will need to develop landscapes that are reconciled to human use and that have enough resilience to bounce back from human and natural disasters. As the Sacramento River riparian restoration project and Guanacaste National Park demonstrate, reconciled landscapes will probably be better for both native organisms and humans.

FURTHER READING

Grunwald, M. 2006. The swamp: the Everglades, Florida, and the politics of paradise. Simon and Schuster. New York. *A history of attempts to protect the everglades.*

Jordan, W. R., M. E. Gilpin, and J. D. Aber, eds. 1987. Restoration ecology: a synthetic approach to ecological research. Cambridge University Press. Cambridge. *This book explores the ecological concepts and ideas involved in the practice of restoration; it was one of the first textbooks in the field.*

Rosenzweig, M. 2002. Win-win ecology. Oxford University Press. Oxford. *An upbeat discussion of how we can incorporate wild critters into human-dominated landscapes.*

Sanderson, E. W. 2009. Mannahatta: a natural history of New York City. Abrams. New York. *By matching an eighteenth century map of Manhattan's landscape to the modern cityscape and applying principles of ecology, Sanderson gives readers both a window into the past and an inspiration for green cities and wild places in the future.*

Walker, B., and D. Salt. 2006. Resilience thinking: sustaining ecosystems and people in a changing world. Island Press. Washington, DC. *A small but informative book on how to make real conservation and restoration work.*

Weisman, A. 2007. The world without us. Picador Press. New York. *A very interesting theoretical examination of how nature might rebound if and when human domination of the planet comes to an end.*

Young, T. P. 2000. Restoration ecology and conservation biology. Biological Conservation 92:73–83. *This is the interesting scientific paper mentioned in the early part of this chapter; it is written at a level that almost everyone can understand.*

13

Conservation in Action

Everything we do affects other species around us. We are the dominant creatures on Earth. We can choose to wipe out most of the species on the planet by continuing on our present course of population growth and accelerated resource use, or we can choose to protect our planet's biodiversity and reverse some of the damage we have already done. It is vitally important for all of us to realize that taking on this major shift in direction is in our own best interests as individuals and as a species. Thus, to protect biodiversity and save wild creatures from extinction requires not only positive action but also changes in lifestyle and changes in our general way of thinking (or not thinking). We must learn to understand, take in, and practice the maxim "think globally, act locally," and we must realize that we are bound with all other forms of life into one gigantic wondrous planetary ecosystem. If we as a species fail at this momentous task, then the diversity of life as we know it will forever be changed, and our planet will be a less friendly place for us as well.

Sometimes this task may seem overwhelming. We are facing problems that are so mind-bendingly complex and difficult to solve that,

as individuals, we tend to shrug our shoulders and go on with our lives, hoping that by some magic, the problems will go away. After all, you may argue, quite logically, "I am only one person among 6.5 billion, and my actions therefore mean nothing." Yet there is an inherent flaw in this line of thinking, and the flaw is in believing that you by yourself are powerless. If you want the truth, here it is: *you are mighty beyond what words can express, and the decisions you make in your life could change the world.* One person's actions are important, and individual people do change the course of history. Just look at people such as Mohandas K. Gandhi, Martin Luther King Jr., Rosa Parks, Rachel Carson, and Harvey Milk. They are all examples of individuals who altered the course of history by the example their actions set.

But you do not need to be an exceptional hero to have a huge impact. Think about how many people you have known in your life. If 10 of those people changed their behavior as a result of following your example, and each of those 10 influenced 10 others and onward, then the ripple effect of your influence would be gigantic. Although one drop of water can't even erode an anthill, the sum of individual drops of water can erode granite mountains. The following pages offer a few suggestions of positive actions you can take to help protect biodiversity and, eventually, yourself.

LEARN FROM THE PROS

When we want to know something, our first impulse is to ask a professional, someone who gets paid to know how to do something. So if we want to know how to minimize our negative impact on the planet, we should ask the professionals. But who would the professionals be? In this case, there is a group called the Union of Concerned Scientists (UCS), which looked into the most effective ways to reduce the environmental impact of Americans' economic lifestyle and published a book describing their findings.

The last few decades have seen a proliferation of these types of books, which list 50, 100, or even 1,000 things that consumers can do to help the planet. Unfortunately, these books have produced disagreement among scientists and consumers alike. Is it really true that choosing paper over plastic at the supermarket would have a positive impact? How about cloth vs. plastic diapers, or styrofoam vs. paper cups for your coffee? Many people questioned whether any of these minor choices made any difference in the long run.

The UCS decided to directly tackle the question of what an environmentally conscious consumer should do by identifying the biggest environmental problems caused by consumer activities. In true scientific fashion, they gathered numerical data about environmental problems and used statistical analysis to determine that the biggest environmental problems related to our everyday consumer activities are global warming, air pollution, habitat alteration, and water pollution. Once they determined these, the UCS team examined how much environmental damage is caused by our everyday life. This allowed them to determine which common activities and practices in the average U.S. household are the most damaging to the environment.

Their results were impressive in their simplicity. Overwhelmingly, the numbers showed that the worst consumer activities that we engage in are (1) driving our cars; (2) eating high on the food chain (i.e., eating meat); (3) eating commercially produced (i.e., with pesticides and fertilizers) fruits, vegetables, and grains; and (4) heating and lighting our houses. Some of these activities are obvious. We all know that driving our cars has a negative impact on the planet, but we may not have known that *driving is by far the worst thing we personally do that harms biodiversity.* We also may not have known that eating meat (from the second or third trophic level) is also detrimental to the planet, or that buying conventionally raised fruits and vegetables is having an impact. These seemingly normal, mundane, "everybody-does-it" activities are far and away having the biggest impact on the planet's ecosystems. Issues such as choosing paper over plastic bags have almost no impact when stacked up against these heavy hitters.

Given the simple list above, it becomes relatively easy to see how one person can have the most positive impact on protecting the world's biodiversity. Here is a list of suggestions based on the UCS's research:

Drive as little as possible or not at all; walk or ride a bicycle as much as possible.

Use public transportation as much as possible.

Choose a place to live that reduces the need to drive.

If you must drive, choose the most fuel efficient, low-polluting vehicle you can afford to buy.

Set goals for reducing your travel.

Eat from lower on the food chain, which means more vegetables and whole grains and less meat (especially beef) and fish.

Buy certified organic produce and products whenever possible.

Buy locally grown products whenever possible (e.g., from farmers' markets).

FIGURE 13.1. A huge variety of delicious organic produce is available at the Saturday morning Embarcadero Farmers' Market in San Francisco, California. Many cities and towns now have farmers' markets, where local produce and agricultural products are for sale directly from the producers. (Photo by M. P. Marchetti.)

Plant a garden and raise some of your own food.

Choose your home to have features that reduce energy expenditure (e.g., smaller size).

Install the most energy efficient appliances and lighting sources you can afford.

Turn your thermostat down in the winter and up in the summer.

Choose an electricity supplier offering renewable energy.

Install renewable energy sources (solar or wind) when possible.

Some of these suggestions may be hard to do at first, but many of them have more than one positive benefit. Let's look at the case, for example, for buying locally grown and organic food (Figure 13.1). Did you know that in the United States, on average, the food you buy at a supermarket travels more than 1,500 miles before it gets to you? This means that for every bag of chips, can of soda, or pineapple you buy, more energy (in terms of fossil fuel used) is put into

transporting the product to where you live than there is energy (in terms of calories) in the food itself! Buying locally grown produce slashes this energy expenditure as well as reducing the amount of CO_2 dumped into the atmosphere. But there are other important benefits of buying locally produced meat and produce. One is that the money you spend on locally grown food stays in the community where you live and does not go to support a super-wealthy shareholder of a giant supermarket chain who lives in a $5 million home in midtown Manhattan. Instead, the money you spend goes directly to the farmer who grew the produce. This allows you to support your local community and provide a livelihood to the people who are your friends and neighbors.

In addition, buying local means that you also have the opportunity to ask questions of the people who grow your food. Which variety tastes the best? Which fruit is the ripest? Was it grown with pesticides or hormones? Locally grown produce is also always fresh and in season. One of the benefits of buying and eating local foods is that the tremendous quantity of energy it takes to get you a ripe tomato in January (typically flown in from the southern hemisphere, thousands of miles away) is not used. Another aspect of the local foods movement is to restore what Amy Trubek calls "the taste of place," an appreciation for the distinctive flavors of food grown in your area and for locally based regional cuisine. This type of thinking has also led to the explosive growth of something called the "Slow Food" movement. "Slow foodies" envision a future food system that is based on the principles of high quality and taste, environmental sustainability, and social justice. They believe there is goodness and value in knowing where and how your food is raised and in preserving our ethnic and regional culinary traditions, things you can get by buying your food locally. Perhaps the most satisfying benefit of buying local from somewhere such as a farmers' market or food co-op is that you get to engage with other people like yourself in a social setting and become more fully integrated into your local community. Farmers' markets, food co-ops, and the like are generally fun places to shop, socialize, and meet people.

For almost all the suggestions listed above, there are similar groups of additional benefits besides helping protect biodiversity. For example, riding your bike every day not only saves fossil fuels but also gives you daily exercise and may make you healthier and perhaps help you to live longer. If a large fraction of us are able to implement some or all of these suggestions, then we can have a positive impact on global environmental issues in a short period of time.

SLOW DOWN

One of the dominant features of our culture is our obsession with "saving time," as though time were something that could be stored in a deep freeze or bank vault. We consume enormous quantities of energy by using "time-saving" gadgets, from dishwashers to power lawn movers to garbage disposals. We drive powerful automobiles at speeds faster than the law allows in order to travel to places as quickly as possible. We eat foods in which there is more energy tied up in the packaging than there is in the food itself (e.g., Kraft Lunchables). All too often, the time "saved" is used for trivial things: to watch TV programs or play video games.

As individuals, we need to consider the environmental cost of all this collective time saving and act accordingly. Plan long trips for more leisurely driving and get better gas mileage. Be willing to take the time needed to use public transportation or car pools. Make chores into social activities. Take the few extra minutes needed to mow your lawn with a hand push mower (and the good, quiet exercise it provides). We are not suggesting a return to living styles of 200 years ago, just some minor but real adjustments to present life styles that might reduce such things as air pollution (and the associated atmospheric warming), the demand for dams

that destroy streams, and the amount of habitat covered up by garbage.

LIVE WITH BLEMISHES

Neatness is the enemy of wildlife. Much traditional landscaping, for example, is open and neatly trimmed, with little room for birds and other animals, and it often requires heavy use of fertilizers and pesticides. Instead let the weeds and bushes grow. Plant native trees and flowers. Our demand for unblemished fruit and vegetables demands the heavy use of pesticides and forces farmers to go to great lengths to control birds and other "pests." Blemished or slightly damaged fruit is still edible. One of our fathers (PBM) had the habit of never eating an apple without taking out his pocketknife and cutting it up. This habit was ingrained in him from being brought up on a farm in the days before the heavy use of pesticides, when finding a small worm in the center of the apple was likely. Adopting simple habits such as this can help to save wildlife (and maybe your own health).

A particularly egregious example of the neatness problem is the manicured suburban lawn. Somehow, it has become desirable in our culture to strive toward a perfect lawn that looks more like an outdoor carpet than something alive. Unfortunately, the perfect lawn requires heavy doses of pesticides, herbicides, and fertilizers to keep the monoculture of grass going. Then, when the grass is ready to be harvested, we cut it with a noisy, gas-guzzling mower and throw the harvest away (often neatly wrapped in a plastic bag), clogging our landfills. Grass from lawns is apparently the single biggest "crop" in the United States, and we throw it away! Meanwhile, the excess fertilizer and pesticides wind up in local streams, ponds, or lakes via storm drains, where they have toxic effects on fish, ducks, and other aquatic life.

Part of a solution to this is to think of your yard and neighborhood as an ecosystem and strive for diversity: appreciate the variety of flowers and grasses that push their way through the dominant turf grass; enjoy the insects that crawl in it or fly over it and create a chorus of sound at night. Despite the warnings of the lawn care industry, it is possible to have a patch of green to sit and play on without swamping it with nasty substances. You can mow your lawn with a hand mower and let the clippings become reincorporated into the lawn. Another part of the solution is to reduce lawn area as much as possible, replacing the grass with low-demand ornamental plants or a vegetable garden.

REDUCE, REUSE, AND RECYCLE

"Reduce, reuse, and recycle" is a slogan that goes well with "Think globally, act locally." We apologize for presenting this overused slogan, but it does have a great element of truth in it. All three general activities can lower your personal contribution to environmental degradation.

Reduce the amount of materials and energy you consume by buying fewer prepackaged goods, driving in an efficient manner (slower, and without jackrabbit starts, etc.), sharing DVDs and books, minimizing the use of heating and air conditioning, and taking other similar actions.

Reuse items as much as you can. Many "disposable" items are reusable, especially containers. For example, if you make homemade wine or beer, you can use old wine bottles year after year. Remember to shop using cloth bags or by bringing "used" bags with you to carry your purchases home.

Recycling is another easy way to reduce your environmental impact, especially in communities with curbside recycling programs. Start the practice, and it soon becomes a habit instead of a chore. Recycling paper, aluminum, and bottles is so easy, in fact, that it is your responsibility to recycle. If your apartment complex or workplace does not have bins for recycling, demand that some be installed. If you have a choice, avoid using materials that cannot be recycled.

CONTROL YOUR PETS

Many people who keep pets tend to think of them as being more human than animal. Because of this, the goods and services provided to pets contribute their share to general environmental degradation. But pets also create special problems in relation to maintaining natural ecosystems.

CATS There are far more cats (top predators) in this world than natural ecosystems could ever accommodate. This is especially true for feral cats, abandoned pet cats, and house cats that spend large amounts of time stalking wild birds. Cats are natural predators, even cats that are stuffed daily with cat food. Although pet cats sometimes kill rats and mice, they rarely have much effect on rodent populations because of the rodents' rapid powers of reproduction (i.e., rodents are r-selected species). We are just beginning to appreciate, however, the numbers of birds that cats kill each year. For example, the pet cat of a lighthouse keeper on tiny Stephen Island off the New Zealand coast wiped out a species of wren single-handedly. Recent studies of cat predation have shown that it is not unusual for a pet cat to kill 200–400 birds per year! Most of the birds are native migrants that have not evolved to handle the artificially high density of predators we have created in our backyards. Such migrants are already in trouble due to destruction of nesting habitats and wintering habitats, so the cats are an added cause of their decline. Imagine the experience of a juvenile white crowned sparrow, reared in California's Sierra Nevada, that follows its parents into low-elevation winter habitat. The inexperienced bird suddenly finds itself in the gardens and backyards of the city of Sacramento, trying to survive in an area with an extraordinarily high density of predators. These are mostly well fed cats with nothing else to do but stalk inexperienced birds. The likelihood of this song bird surviving the winter to return to its mountain home is correspondingly low.

The high place cats hold in society was demonstrated in the city of Davis, California, in 2004. A couple of coyotes started foraging in a city park, and one of their favorite prey became household cats that had been let out to play with the birds. Not surprisingly, the city quickly dispatched the errant coyotes. If they had been left alone, they would have greatly reduced cat densities in the park, and eventually there likely would have been an increase in nesting and migratory birds. This effect has been demonstrated in southern California, where natural canyons between housing tracts that have coyotes doing cat control have much more diverse bird faunas than those without coyotes. It is interesting to speculate how much more diverse our local parks might be if we either encouraged coyotes or required cat owners to keep their cats indoors.

If you own a cat, here are some things you can do that can help minimize its impact.

1. Keep it indoors, especially during fall and spring when migratory birds are most abundant. This is not as cruel you may think because cats spend much time sleeping anyway. Cats kept indoors from the time they are small kittens are typically well adjusted to their "habitat."

2. When you want to adopt a cat, obtain a kitten from someone who has an indoor cat, because hunting skills are learned in part from the mother. Keep the kitten indoors as much as possible for the first year (and thereafter).

3. Have your cat spayed or neutered.

4. If you leave an area and cannot take your cat with you, do not just turn it loose; find someone to adopt it or else take it to the local SPCA.

5. Do not feed abandoned cats unless you plan to adopt them. The best thing to do is to capture them and turn them over to an animal shelter; otherwise, you will be helping to increase the density of healthy predators on birds. The feeding of cats in public parks is particularly harmful.

6. Play with your cat. If it looks bored, it is your responsibility to entertain it.

Both authors have lived with cats and therefore recognize the delight of having an occasionally affectionate, semi-wild animal in our lives. But in today's world, keeping a cat entails responsibility.

DOGS Dogs, like cats, are predators, although they are much more domesticated. However, if you live out in the country and let your dog run loose at night, you are probably contributing to the loss of wildlife, especially deer which can be chased to exhaustion (or into the path of a car). Often, coyotes are blamed for killing livestock that were actually killed by loose dogs. An equally serious, but less appreciated, problem is the effect dogs have on wildlife when let off their leashes to "explore" a wild area. Anyone who has kept a dog and let it do this knows how much most dogs enjoy plunging through the underbrush, poking their noses into holes, flushing birds, and chasing rabbits and squirrels. What is harmless fun for your dog may be deadly serious for the wildlife it disturbs, however, making the animals more vulnerable to real predators, disturbing nesting or care of young, or reducing their ability to forage effectively. Wild areas that are heavily frequented by people with dogs consequently contain less wildlife.

One obvious thing to do in this regard is to keep your dog on a leash when in a natural area, especially if signs tell you to do so! Try to exercise your dog as much as possible in non-wild areas, preferably in park areas designated for that purpose. If you have a dog, you should work and donate money toward the establishment of "dog parks," where dogs and their owners can run free together. Such actions will also reduce the probability that foxes, coyotes, and other animals will catch a disease your pet carries (and vice versa).

BIRDS AND REPTILES The main problems with pet birds and reptiles from a wildlife perspective are (1) the demand for exotic species caught in the wild and (2) the escape of species likely to become pests in the wild or competitors with native species. These problems are easy to solve:

• Buy only animals that were bred in captivity.

• Make sure you have a secure place to keep your pets.

• Don't release any animals into the wild.

An example of unexpected consequences of keeping wild-type pets is illustrated by the keeping of turtles. Most turtles kept as pets in North America are red-eared sliders, native to the eastern United States. These turtles are frequently released into the wild when their keepers, usually parents of small children, tire of them. In the wild, they compete with and spread diseases to native turtles, which are becoming increasingly scarce in many areas. Sometimes, even the release of native reptiles back into the wild can create problems. One of the reasons the desert tortoise of southern California is endangered is that people who have kept them as pets have returned them to the wild carrying a disease that then spreads to the wild tortoises.

FISH The problems with aquarium fish are similar to those of birds and reptiles, but the people who keep them feel less constrained about letting them go when they are tired of them. This is the reason that goldfish are one of the most introduced species of fish and that sewage treatment plants often support large populations of guppies, flushed to their freedom. Aquarium fishes are a particular problem in small isolated waters such as desert springs, where they have contributed to the extinction or endangerment of a number of species of fish and invertebrates. In most cases, aquarium fishes released to the wild do not make it; they either get eaten by a predator, or they die when the water gets too cold or too hot. However, even in this situation, they can spread potentially devastating diseases to wild fish. If you want to get rid of pet fish, the responsible thing to do is to either find someone else to take them or, if that fails, kill them.

If you keep saltwater fishes, you should not buy any unless the dealer can document where the fish came from and how they were captured. Most saltwater aquarium fishes are taken from the wild, although cultured fish are becoming more available. Many of the most attractive saltwater fishes are the inhabitants of coral reefs. Collection techniques can be highly destructive of the reefs, such as the use of sodium cyanide (a poison) to knock out fish, dynamite to stun fish, or rocks dragged on ropes to herd the fish into nets, which crushes living corals. Such techniques kill enormous numbers of fish for the few that wind up in the aquarium stores, and it destroys their habitat as well.

REDUCE OFF-ROAD VEHICLE USE

Off-road vehicles (ORVs) have become enormously popular and are a tribute to human mechanical ingenuity and to the abundance of leisure time available to Americans. They include four-wheel drive trucks and cars, a variety of motorcycles, special "all-terrain" vehicles, dune buggies, snowmobiles, and, most recently, mountain bikes. The use of off-road vehicles, especially on public lands, has been expanding exponentially, thanks in good part to campaigns by the vehicle manufacturers. The agencies managing public lands have by and large been unable to cope with the invasion of ORVs, often because they lack the person power to regulate ORV use or assess the damage these vehicles cause.

It should not come as a surprise that ORVs cause environmental damage and degrade local ecosystems. The most obvious effect of ORVs is on steep hillsides, where a single motorcycle track can begin the process of erosion and can eventually produce a large gully. The erosion not only eliminates the hillside plants upon which wildlife depend but also causes sediments to fill in streams, thereby reducing aquatic life, including fish. Deserts are particularly vulnerable to harm from ORVs because of their accessibility and fragile biotic communities. A "worthless bush" smashed by an ORV may be a creosote bush that took hundreds of years to grow and which shelters birds, kangaroo rats, and desert iguanas.

In the Mojave Desert of Southern California, a major conflict has arisen between those seeking to protect the endangered desert tortoise and those who use ORVs. ORVs unwittingly smash tortoises and their habitat. ORV users take tortoises home as souvenir pets, and they often leave their garbage behind in the desert.

The latter act is a problem because it has caused a population explosion in scavenging ravens, which are also major predators on desert tortoise babies.

Snowmobiles are another form of ORV that can cause harm. There is often a surprisingly large amount of animal life under the snow pack, which is insulating the ground from extreme cold. Small rodents have runways in the snow that allow them to get to food sources; snowmobiles smash the runways, and the rodent populations may be reduced as a result. This, in turn, may affect the populations of their predators, such as hawks and owls.

Noise is yet another major problem in ORVs, because each vehicle carries with it a dome of noise that may extend a kilometer in all directions. This not only diminishes the aesthetic value of a wild place for people who are seeking quiet and solitude, but also disturbs wildlife, altering behavior patterns. Bighorn sheep and elk, for example, may forage less efficiently if scared out of prime feeding areas.

The best solution to the ORV problem is not to use them in wild areas. Recreational use should be confined to special parks created from gravel pits and to other areas already destroyed by humans. If you are an ORV user, you should do the following:

1. Stay on roads or ORV trails.
2. Stay out of wilderness areas or other areas where ORVs are banned.
3. Make your vehicle as quiet as possible.
4. Take your garbage home with you.
5. Respect wildlife habitat.
6. Get involved in volunteer actions to restore habitats degraded by ORVs and other activities.

Do not do anything with your vehicle without first thinking of the long-term consequences. For example, if you drive across the countryside in a "new" area, you will be creating a trail that others are likely to follow. Your impact may be minimal, but the cumulative impact of those that follow could be disastrous.

Another alternative to motorized ORV use is use of mountain bicycles, which can provide many of the same thrills and spills in wild areas without noise or severe environmental damage. Even mountain bikes have to be used with discretion, however, because heavy use of a trail can create problems similar to those created by motorcycles. Damage is particularly likely in natural areas close to urban areas.

REDUCE POWER-BOATING

Modern powerboats can be every bit as harmful as off-road vehicles. Their wakes accelerate erosion of stream banks and lake shores and disturb nesting birds such as grebes, which build floating nests in beds of rushes and cattails. Powerboats pollute the water and air with gasoline and oil. Their noise and speed make them incompatible with wildlife. Large natural lakes with heavy use by powerboats are devoid of waterfowl or have greatly reduced populations. The incredible noise that many boat engines make can disturb wildlife and the tranquility that many people seek when coming to a lake or stream. Particularly obnoxious are jet skis and other "personal watercraft," which can go into shallow water (often a refuge for waterfowl), make very loud noises, and pollute the water with their inefficient two-cycle engines. In recent years, major improvements have been made to outboard engines to reduce the pollution and noise they create, a positive trend. This reduces some of the problems power-boating causes, but it does not eliminate them.

A potential solution to the powerboat problem is to restrict large-horsepower or noisy or fast boats to selected reservoirs, lakes, and rivers. The preferred methods of boating in natural areas should be sailboats, sailboards, canoes, kayaks, and other quiet, non-polluting vehicles. Keep in mind, however, that there can be too much of a good thing; a river crowded with canoes and kayaks may also have its wildlife populations diminished through constant disturbance.

BE A THOUGHTFUL ECOTOURIST

Is it possible to love nature to death? The growing crowds in our parks and natural areas are telling us the answer may be yes. Even places as remote as the frigidly cold Antarctic or the incredibly dry deserts of Namibia are suddenly being visited by large numbers of tourists (Figure 13.2). There is concern that all these visitors are disrupting the local wildlife, which is usually the focus of the visits. In the popular national parks of Kenya, vehicles full of tourists are so common that predatory animals may use them as cover when stalking their prey. These are signs that the people coming to see animals in the wild are changing the behavior of these animals. One of the reasons for this problem is that people often go into a wild area with expectations of seeing the dramatic events shown in wildlife specials on television. At the very least, they want to get close enough to some spectacular animal to get a dramatic photograph as a souvenir of the trip. Efforts to see or photograph wildlife close up in the short time available on a vacation trip requires intruding on the wildlife, often at times when the animals are resting, breeding, or taking care of young. As the California Department of Fish and Game points out in one of its brochures for ecotourists, "There's a fine line between viewing and victimizing wildlife."

Of course, the other side of this issue is that ecotourists spend lots of money, making important contributions to the economies of impoverished nations and rural areas of North America, and they often gain an increased appreciation for conservation during their travels. Many popular natural areas would not exist without these tourist dollars or at least would not be managed as well. A solution to the problem is managing the behavior of ecotourists to minimize their effects on wildlife. If you are visiting a natural area, here are some things you can do to reduce your impact:

1. Carry a good pair of binoculars so you do not have to get so close to observe birds and mammals.

FIGURE 13.2. The impact of ecotourism can be felt in even the most remote locations on the planet. This caravan of tourists is in Sossusvlei, a remarkable area of huge brick-red sand dunes located in the central Namib Desert (one of the driest locations on the planet), in Namibia. (Photo by M. P. Marchetti.)

2. Stay on the designated roads and trails; use blinds and observation platforms if available.

3. Be patient, be quiet, and move slowly near wildlife. Spend time watching individual animals rather than quickly moving once you have added a species to your list of things seen.

4. Spend time observing the little things, not just the "charismatic megafauna." Notice the smaller organisms such as flowers, insects, or mushrooms, which can be especially good subjects for photography (Figure 13.3). Become a butterfly and bug watcher. What they do is every bit as dramatic as what happens with birds and mammals, just on a smaller scale.

5. Go out with a guide or friend who knows an area. This will increase your probability of seeing wildlife and reduce your impact on it if your guide is sensitive.

6. Be willing to go out when the animals are out, which is often in the evening and early morning, rather than disturbing them during the day so you can see them at your convenience.

7. If you are driving through a wildlife area, stay in the car as much as possible. It can

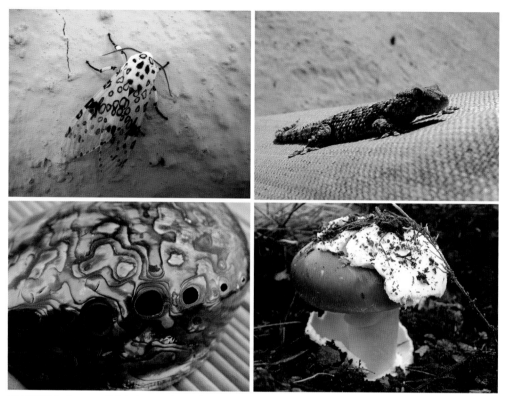

FIGURE 13.3. Amateur digital photography can be a creative and fascinating way to interact with the natural world that has a low impact on natural ecosystems. These four pictures were all taken with an inexpensive "point and shoot" camera by one of the authors, and they are great nondestructive souvenirs of a trip to wild places. (Photos by M. P. Marchetti.)

serve as a movable blind, so you are likely to see more natural behavior.

8. Nature preserves and wildlife areas are notoriously short of funds for management, so donate money to those you use. This is especially true in developing countries.

EAT FISH WITH CARE

The capture of wild fish for consumption and sport is a major recreational activity and a significant source of food for many people. Fisheries range from subsistence fishing, still practiced in a few areas of the world, to sport fishing, to commercial fishing, often on an industrial scale. Unfortunately, fishing, especially commercial fishing, is having a major impact not only on populations of harvested species but also on the ecosystems they inhabit. Dragging

trawl nets across the bottom of the ocean to capture fish, for example, has been likened to the clear-cutting of forests, where the consequences to oceanic life are similarly damaging. This is an area where you can make a difference starting with your eating habits.

EAT ONLY MARINE FISH THAT ARE SUSTAINABLY HARVESTED Preferably, eat only those certified as such by the Marine Stewardship Council. These species are fewer all the time, so you need to check carefully on websites of the Monterey Bay Aquarium, Audubon Society, or Blue Ocean Institute to find species that you can consume with minimal impact.

EAT FISH FROM SUSTAINABLE FRESHWATER AQUACULTURE OPERATIONS Such operations feed fish mainly vegetable-derived foods, rather than ground up marine fish. Examples include tilapia and channel catfish. Try to buy locally grown fish as much as possible; much tilapia sold in the United States, unfortunately, is grown under unknown conditions in poor countries.

HARVEST YOUR OWN FISH   In the United States, we have many freshwater species that can sustain considerable harvest, ranging from sunfish and bass to various carp species. Various Asian carps have such large populations that they are considered a nuisance in many rivers, yet they are potentially an enormous source of fish for markets. The reason they are not harvested for food is mainly cultural: Americans are not used to eating bland, somewhat bony fish, even though the rest of the world thinks they are wonderful eating (think of steamed fish with tasty sauces). A secondary problem with wild caught freshwater fish is that their flesh is often contaminated with pesticides, heavy metals, and other pollutants. Most local areas publish guides to tell you which fish are most contaminated and how often you can eat local fish without harm.

REDUCE WATER CONSUMPTION

There is an old saying in the American West that "Whiskey is for drinking, and water is for fighting." As Marc Reisner eloquently describes in his book *Cadillac Desert,* battles over water make up much of the history of the American West. Until recently, the battles have been mainly among human users of the water supplies, most dramatically between cities and farmers. Los Angeles grew, for example, partly at the expense of farms in the Owens Valley, when the Owens River was captured and sent west through an aqueduct. Unfortunately, the really big losers in the water wars have been the streams and rivers, along with their associated fish and wildlife. Salmon runs have collapsed in the Central Valley of California from one to two million per year to a few thousand. Freshwater fishes in California are going extinct at a rate of about one species or subspecies every five years. The majority of endangered species of wildlife depend on riparian forests, which depend on flowing rivers.

Increasingly, it is being recognized that the problem with water is not so much that there is not enough water but that so much of it is used inefficiently. Urban areas throughout the world have been forced to recognize this fact during recent droughts and have found ways to reduce water use by astonishing amounts, up to 40 percent in some cases. Some of the biggest savings in water, however, can come through better management of water used in agriculture. Agricultural economists have shown that relatively small changes in irrigation practices can yield large savings of water. Even bigger savings could be obtained if the acreage of some water-intensive crops, such as alfalfa, pasture, and cotton, were reduced, especially acreage that is irrigated by flooding it with water. Presumably, much of the water resulting from improved agricultural practices could be used to restore aquatic environments and valuable fisheries.

Changes such as those described above (in a greatly oversimplified fashion) will not come easily, no matter how much sense they seem to make. The reason, of course, is that water reform costs a great deal of money and forces changes in traditional ways of doing business. The only way such change is going to take place is through the political process. Becoming involved in, or at least aware of, water politics will clearly help resolve one of the great political issues of our time. Water issues are important at all levels of human organization, from local to international. As activists on the Hudson River, New York, have discovered, restoring a small headwater stream not only has tangible local results but also improves the quality of drinking water and helps to change public attitudes all the way downstream to the mouth of the river.

In the meantime, you can raise your own environmental awareness by being conservative in your personal use of water: use low-flow showerheads, take shorter showers, install low-flush toilets, use drip irrigation in your garden, water trees with water first used for other purposes, don't wash your car as often, plant drought-resistant ornamental plants, and take other water-conscious actions. During a recent drought in California, urban users showed an astonishing ability to reduce their water use by 30 to 40 percent. In Brisbane, Australia, severe drought in 2008 restricted water use to 35 gallons per person per day, pretty much excluding the use of water for landscaping. The need for such savings can no longer be dismissed as something unique to dry areas; even urban areas in the relatively wet areas of the country experience water shortages at times. Climate change is likely to make water shortages even more frequent; therefore, we should live like we are always in a drought.

You can also consume less water by changing your diet to avoid eating meat, especially red meat. Producing one pound of beef apparently takes about 2,500–5,200 gallons of water, depending on where the animal lives. Most of this water is used not for the cows themselves but in growing the cereal crops to feed them. According to Ed Ayres of the Worldwatch Institute, if you pass up a single hamburger, you'll save as much water as you would from taking 40 showers with a low-flow shower head. The numbers quoted about how much water it takes to produce a pound of any kind of meat actually vary widely (e.g., the beef industry estimates it takes 400–800 gallons per pound of beef), but from a water point of view, grass-fed cattle on non-irrigated pastures consume significantly less water than do cattle fed conventionally grown grains and alfalfa. In any case, it make sense that if you eat lower on the food chain (e.g., if you eat plants), then you will reduce the demand for water.

JOIN AN ENVIRONMENTAL GROUP

The key to protecting the environment that we humans need is to take collective actions that influence the activities of local, state, and federal governments. Collective action requires organization, and this involves individuals joining together in environmental organizations. Much of the environmental protection that exists today is the direct result of lobbying and other activities of environmental organizations. There are groups to fit nearly every political viewpoint and

need, from organizations supported largely by industry (e.g., Keep America Beautiful, which is supported by the container industry, the chief source of litter), to mainstream groups such as the Wilderness Society, the Sierra Club, and the Nature Conservancy, to the radical Earth First! group. What follows is a very incomplete list of organizations that have a strong presence. If you volunteer for one of them or donate money, you are likely to see the results in your own backyard. All such organizations have websites that you can check for updated information. Important organizations that are not included on this list are the many local groups set up to involve communities in conservation, from protecting watersheds to restoring local parks. As the authors know from their own involvement in local organizations, there is incredible satisfaction in seeing an area transformed from a degraded eyesore to species-rich natural area that is the pride of a local community (e.g., Putah Creek corridor in Yolo and Solano Counties, California).

AUDUBON SOCIETY This is a large environmental group with many local chapters and a wide interest in a range of environmental matters. Typically, local chapters have monthly meetings, a newsletter, and field trips (mainly for birding). In some places, they also manage hawk or owl sanctuaries. Volunteers are always needed to help maintain the society's high level of local activism. The National Audubon Society publishes one of the best environmental magazines and works on many national environmental issues.

THE NATURE CONSERVANCY (TNC) This national organization has been extraordinarily effective in acquiring natural areas by direct action: buying land or conservation easements or acquiring them through donation. It has preserves established all over the world and is always looking for volunteers, interns, and, occasionally, paid employees.

SIERRA CLUB This is an environmental group that is nearly a century old. It makes good use of volunteers in its educational and environmental efforts.

ANGLING ORGANIZATIONS Anglers have been getting organized in the last few years because water diversions and pollution have reduced fishing opportunities. Their efforts to protect streams have paid off in terms of many broader issues such as saving riparian forests and other habitat for wildlife. If you like to fish, you should join one of these organizations.

WATERFOWL ORGANIZATIONS Duck hunters have a good record of raising money and using their clout to protect wetlands, necessary for the birds they love to hunt. Ducks Unlimited and similar local organizations can always use members, and occasionally interns.

LEGAL ORGANIZATIONS Many environmental lawsuits are handled by the Sierra Club Legal Defense Fund, the Environmental Defense, and the Natural Resources Defense Council. They mainly need your money to keep going, but they can sometimes use interns and volunteers.

DEFENDERS OF WILDLIFE, THE WILDERNESS SOCIETY, AND SIMILAR ORGANIZATIONS These are national organizations, often with state offices, that have been particularly effective in lobbying national and state governments on environmental issues.

BE AWARE OF YOUR BIOREGION

Joining local environmental organizations can be the first step toward living in a region as if you really belong there. We are a highly transient culture and think little about uprooting ourselves and moving long distances for a new job, to go to school, or for other life events. The problem with this pattern is the loss of an individual's identification with a local place. People who strongly identify with and know the place in which they live are more likely to defend it against unfavorable changes. This is the basis of bioregionalism. This philosophical approach has been defined by bioregional guru Robert L. Thayer as "grassroots on-the-ground action toward the resolution of environmental and social issues by voluntary non-profit groups that identify strongly with natural regions and local communities." This kind of action results in

such things as protection or restoration of local wildlife or natural areas, support (through purchase power) of local organic farmers' products, voting to increase taxes to buy farmland as open space, and creation of art and writing celebrating the local landscape.

What is a bioregion? A bioregion, as defined by Thayer, is a geographic location in which there are "distinct communities of life, both human and nonhuman, where implicit conditions suggest particular adaptations." This definition has three main points: (1) humans are part of the ecosystem, (2) humans have strong interactions with the local environment, and (3) each area has its own distinct characteristics, from its climate to its plants and animals, and requires local knowledge to make it work in a sustainable fashion. Often, bioregions can be defined by watershed boundaries, such as creeks, rivers, or lakes. When we understand where our water comes from and goes to, when and where our local food is grown, how seasonal patterns effect wildlife, and how literature and art produced in a region is influenced by the natural environment, then we can get a better feeling for what it means to really inhabit a location. This is the stuff that makes bioregionalism so fascinating and can help us to understand why a lifestyle that reduces our personal demand for resources can have a major impact locally.

VOTE!

In the United States, we have the privilege of living in a relatively free democratic society. We do not live under a dictatorship, an oligarchy, a monarchy, or a theocracy, or even in a socialist state. Because we live in a democracy, we generally get to have a say in how we are governed. Many of our direct ancestors gave enormous amounts of time, energy, and even their lives to make sure that we have this right and to ensure that this right is granted to every member of our society. As a result, the people of the United States have a responsibility and an obligation to vote on Election Day. Yet you often hear, or may have uttered yourself, some of the following types of statements:

> "My one vote doesn't count for anything."

> "The elections these days are all about media."

> "Elections are probably rigged, and it doesn't matter who you vote for."

> "Both candidates are basically two sides of the same coin, and I don't like either one."

> "I'm just too busy to vote."

All of these excuses are difficult to listen to. We have had a series of very close and much contested presidential and federal elections over the last few decades in the United States (e.g., Al Franken won his senate seat in Minnesota by little more than 300 votes out of the 2.8 million people who voted!). Yet if you look at the numbers since 1960, the average voter turnout has been less than 50 percent. This means that less than half of the population actually bothered to cast a vote. When you think about these numbers and realize that most elections are fairly close, this suggests that less than one-quarter of the U.S. population determined the future for the other three-quarters in each of the last four decades! This is outrageous, when you stop and think about it. No wonder many people feel disenfranchised and underrepresented; in fact, they are not casting their votes, and their opinions are therefore not being heard.

Even if you have some issues with the big federal election process, it is important to stop and consider all the local elections and politics. Do you realize that a huge amount of legislation that helps protect natural areas, biodiversity, and wildlife occurs at the local level? How do you think a city or county decides where to build a new subdivision or a new shopping center, or what to do with its sewage or garbage, or where and how to construct new parks or recreation areas? We elect local officials to handle these questions and to represent our interests. This means we choose people we feel will vote on issues in a way that aligns with our interests and values. Do you know that many local elections are won by a margin of only a few votes? In a close election, by convincing your close friends to vote a certain way, you can actually play a big role in how the local government is run, how resources are distributed, and how biodiversity is going to be impacted. That is real power, and power is the fuel that drives change. You can make a difference, and you can effect change. Voting is one of the best ways to empower ourselves and our communities.

VOTE WITH YOUR MONEY

In addition to voting on Election Day, keep in mind that every time you step up to a cash register to pay for something, you are in effect voting with your money. You are telling someone (a marketing executive, a company, a local government) that you feel that the product you are buying is important enough for you to spend money on. This means that if you purchase goods or services that are environmentally friendly or benign, you are telling the economy that you value those goods or services. We essentially vote every single day whenever we spend our money. So when you next spend your money, vote wisely. Think about how your dollars are going to impact wildlife.

CONCLUSION

Regardless of what you do or how you choose to live your life, one thing is true: the more you know and the more you understand, the better you will be able to make decisions. The truth is that we make hundreds of decisions that affect nature and biodiversity every single day, and most of the time, we don't even realize it. These decisions have power and impact, and the best thing we can do is to be aware of our choices and the effects they have. A general intellectual strategy is to do the following:

- Be informed about the effects of your daily actions.

- Read widely, but with a skeptical mind.

- Share your knowledge with others.
- Vote wisely with your money and time.
- Turn your thoughts and words into actions that will benefit the planet.

Knowledge is power, and power can effect change. Ultimately it's all in your hands.

> If you choose not to be consciously involved in the conservation of forms of life other than your own, you should at least be aware that by doing nothing you are still having an impact on the biota of this planet. The water you drink, the food you eat, the land you live on, and the air you pollute were all obtained at the expense of other creatures. The decisions we make today on how we are going to share these resources will determine which other species will inhabit Earth for the indefinite future.
>
> R.L. Thayer, 2003

FURTHER READING

Brower, M., and W. Leon. 1999. The consumer's guide to effective environmental choices: practical advice from the Union of Concerned Scientists. Three Rivers Press. New York. *This really is the best source of information about how much impact we are having in our daily lives and what concrete steps we can take to address these issues.*

Clover, C. 2006. The end of the line: how over fishing is changing the world and what we eat. University of California Press. Berkeley. *A book that makes you reconsider eating marine fish!*

Halweil, B. 2004. Eat here: homegrown pleasures in a global supermarket. W. W. Norton and Co. New York. *An insightful and timely book that indicates how important food, farms, and rural cultures are to the world.*

Kingsolver, B., C. Kingsolver, and S. Hopp. 2007. Animal, vegetable, miracle: a year of food life. Harper Perennial. New York. *Kingsolver is a novelist who turns her biology-trained eye to a year's worth of local eating and production of her own food. Very well written and inspiring.*

Pollan, M. 2008. In defense of food: an eater's manifesto. Penguin Press. New York. *One of the best recent books on the impact our eating and gastronomic habits have on the planet and our health.*

Reisner, M. 1993. Cadillac desert: the American West and its disappearing water. Penguin Press. New York. *This book is an amazing look at the role that water and water politics have played in the American West. There is also a fantastic four-part PBS video series of the same name. You will never look at water the same after reading this.*

Thayer, R. L. 2003. LifePlace: bioregional thought and practice. University of California Press. Berkeley. *A wonderful book that explores what it means to live a truly bioregional life.*

Trubek, A. B. 2008. The taste of place: a cultural journey into terroir. University of California Press, Berkeley. *An exploration of why local food just tastes better.*

BIBLIOGRAPHY

These and many other publications informed and helped illustrate this text.

Balmford, A., et al. 2002. Economic reasons for conserving wild nature. Science 297(5583):950–953.

Barthlott, W., J. Mutke, M. D. Rafiqpoor, G. Kier, and H. Kreft. 2005. Global centers of vascular plant diversity. Nova Acta Leopoldina 92:61–83.

Bayon, R. 2008. Banking on biodiversity. State of the world 2008: innovations for a sustainable economy (pages 123–137). Worldwatch Institute. Washington, DC.

Beard, J. S. 1955. The classification of tropical American vegetation-types. Ecology 36(1):89–100.

Boag, P. T., and P. R. Grant. 1984. The classical case of character release: Darwin's finches (Geospiza) on Isla Daphne Major, Galapagos. Biological Journal of the Linnean Society 22(3):243–287.

Bradshaw A. D. 1987. The reclamation of derelict land and the ecology of ecosystems. In Restoration ecology: a synthetic approach to ecological research. W. R. Jordan III, M. E. Gilpin, and J. D. Aber, eds. Cambridge University Press. Cambridge.

Bright, C. 1998. Life out of bounds: bioinvasion in a borderless world. W. W. Norton and Co. New York.

Brower, M., and W. Leon. 1999. The consumer's guide to effective environmental choices: practical advice from the Union of Concerned Scientists. Three Rivers Press. New York.

Buel, J. W. 1889. The living world: a complete natural history of the world's creatures. Holloway Publishing. St. Louis, MO.

Bulmer, M. G. 1974. A statistical analysis of the 10-year cycle in Canada. The Journal of Animal Ecology 43(3):701–718.

Carson, R. 1962. Silent spring. Houghton Mifflin. Boston.

Carwardine, M., and S. Fry. 2009. Last chance to see. Collins Publishing. New York.

Chapman, J. A., and G. A. Feldhamer, eds. 1982. Wild mammals of North America: biology management economics. Johns Hopkins University Press. Baltimore, MD.

Clausen, J., D. D. Keck, and W. M. Hiesey. 1948. Experimental studies on the nature of species, volume III: environmental responses of climatic races of *Achillea*. Carnegie Institution of Washington Publication 581. Washington, DC.

Clover, C. 2006. The end of the line: how over fishing is changing the world and what we eat. University of California Press. Berkeley.

Costanza, R., et al. 1997. The value of the world's ecosystem services and natural capital. Nature 387:253–260.

Cox, G. W. 1999. Alien species in North America and Hawaii: impacts on natural ecosystems. Island Press. Washington, DC.

Crawford, W. P. 1992. Mariner's weather. W. W. Norton and Co. New York.

Crosby, A. W. 1986. Ecological imperialism: the biological expansion of Europe, 900–1900. Cambridge University Press. Cambridge.

Daily, G. C., ed. 1997. Nature's services: societal dependence on natural ecosystems. Island Press. Washington, DC.

Daly, H., and Farley, J. 2004. Ecological economics: principles and applications. Island Press. Washington, DC.

Darwin, C. 1859. On the origin of species: by means of natural selection, or the preservation of favored races in the struggle for life. Dover Publications. Mineola, NY.

Diamond, J. 1999. Guns, germs and steel. W. W. Norton and Co. New York.

Diamond, J. 2005. Collapse: how societies choose to fail or succeed. Viking Press. New York.

Dombeck, M. P., C. A. Wood, and J. E. Williams. 2003. From conquest to conservation: our public lands legacy. Island Press. Washington, DC.

Dow, K., and T. E. Dowling. 2007. The atlas of global climate change. University of California Press. Berkeley.

Elton, C. S. 1958. The ecology of invasions by animals and plants. University of Chicago Press. Chicago.

Flannery, T. 2001. The eternal frontier: an ecological history of North America and its people. Grove Press. New York.

Forthergill, A., V. Berlowitz, M. Brownlow, H. Cordey, J. Keeling, and M. Linfield. 2006. Planet Earth. University of California Press. Berkeley.

Fowler, C. W. 1990 Density dependence in northern fur seal *(Callorhinus ursinus)*. Marine Mammal Science 6(3):171–195.

Gardner, G. 2002. Invoking the spirit: religion and spirituality in the quest for a sustainable world. Worldwatch Paper 164. Worldwatch Institute. Washington, DC.

Gardner, G., and T. Prugh. 2008. Seeding the sustainable economy. State of the world 2008: innovations for a sustainable economy (pages 3–17). Worldwatch Institute. Washington, DC.

Gause, G. F., N. P. Smaragdova, and A. A. Witt. 1936. Further studies of interaction between predators and prey. The Journal of Animal Ecology 5(1):1–18.

Glantz, M. H. 1981. Consideration of the societal value of an El Niño forecast and the 1972–1973 El Niño. In Resource management and environmental uncertainty: lessons from coastal upwelling fisheries, volume 11 (pages 449–476), in the Wiley series in Advances in Environmental Science and Technology. M. H. Glantz and J. D. Thompson, eds. Wiley Interscience. New York.

Goodman, M., G. W. William Moore, and G. Matsuda. 1975. Darwinian evolution in the genealogy of haemoglobin. Nature 253:603–608.

Gould, S. J. 1994. Hen's teeth and horse's toes: further reflections in natural history. W. W. Norton and Co. New York.

Gould, S. J., and S. Rose. 2007. The richness of life: the essential Stephen Jay Gould. W. W. Norton and Co. New York.

Grant, P. R. 1986. Ecology and evolution of Darwin's finches. Princeton University Press. Princeton, NJ.

Groom, M. J., G. K. Meffe, and C. R. Carroll. 2005. Principles of conservation biology, third edition. Sinauer. Sunderland, MD.

Grunwald, M. 2006. The swamp: the everglades, Florida, and the politics of paradise. Simon and Schuster. New York.

Halweil, B. 2004. Eat here: homegrown pleasures in a global supermarket. W. W. Norton and Co. New York.

Hardin, G. 1968. The tragedy of the commons. Science 162:1243–1248.

Hurlbert, A. H., and J. P. Haskell. 2003. The effect of energy and seasonality on avian species richness and community composition. The American Naturalist 161:83–97.

Jordan, W. R., M. E. Gilpin, and J. D. Aber, eds. 1987. Restoration ecology: a synthetic approach to ecological research. Cambridge University Press. Cambridge.

Kingsolver, B., C. Kingsolver, and S. Hopp. 2007. Animal, vegetable, miracle: a year of food life. Harper Perennial. New York.

Krebs, C. 2008. The ecological world view. University of California Press. Berkeley.

Krebs, J. R., and N. B. Davies. 1987. An introduction to behavioral ecology, second edition. Blackwell Publishing. Oxford.

Kurlansky, M. 1997. Cod: biography of a fish that changed the world. Penguin Books. New York.

Leopold, A. 1990. Sand County almanac. Ballantine Books. New York.

Leslie, P. H. 1957. An analysis of the data for some experiments carried out by Gause with populations of the protozoa, *Paramecium aurelia* and *Paramecium caudatum*. Biometrika 44(3/4):314–327.

Lockwood, J., M. F. Hoopes, and M. P. Marchetti. 2006. Invasion ecology. Blackwell Press. Oxford.

Lott, D. 2002. American bison: a natural history. University of California Press. Berkeley.

Martin, P. 2006. Twilight of the mammoths: ice age extinctions and the re-wilding of America. University of California Press. Berkeley.

Moyle, P. B. 1993. Fish: an enthusiast's guide. University of California Press. Berkeley.

Moyle, P. B. 2002. Inland fishes of California. University of California Press. Berkeley.

Mutke, J., and W. Barthlott. 2005. Patterns of vascular plant diversity at continental to global scales. Biologiske Skrifter 55:521–531.

Naiman, R. 1988. Animal influences on ecosystem dynamics. Bioscience 38(11):750–762.

National Research Council. 1995. Science and the Endangered Species Act. National Academy Press. Washington, DC.

Pavlik, B. M. 2008. The California deserts: an ecological rediscovery. University of California Press. Berkeley.

Perrins, C. M. 1965. Population fluctuations and clutch-size in the great tit, *Parus major* L. The Journal of Animal Ecology 34(3):601–648.

Pianka, E. R. 1988. Evolutionary ecology, fourth edition. Harper and Row Publishing. New York.

Pollan, M. 2008. In defense of food: an eater's manifesto. Penguin Press. New York.

Porter, W. P., J. W. Mitchell, W. A. Beckman, and C. B. DeWitt. 1973. Behavioral implications of mechanistic ecology: thermal and behavioral modeling of desert ectotherms and their microenvironment. Oecologia 13:1–54.

Postel, S. 1996. Dividing the waters: food security, ecosystem health, and the new politics of scarcity. Worldwatch Paper 132. Worldwatch Institute. Washington, DC.

Quammen, D. 1996. Song of the dodo: island biogeography in an age of extinction. Scribner. New York.

Reisner, M. 1993, Cadillac desert: the American West and its disappearing water. Penguin Press. New York.

Rohde, R. A., and R. A. Muller. 2005. Cycles in fossil diversity. Nature 434:208–210.

Romanes, G. J. 1896. Darwin and after Darwin: an exposition of the Darwinian theory and a discussion of post Darwinian questions. The Open Court Publishing Company. Chicago.

Rosenzweig, M. 2002. Win-win ecology. Oxford University Press. Oxford.

Sanderson, E. W. 2009. Mannahatta: a natural history of New York City. Abrams. New York.

Shaw, J. H. 1985. An introduction to wildlife management. McGraw Hill. New York.

Snyder, G. 1990. Practice of the wild. North Point Press. New York.

Stockwell, C. A., A. P. Hendry, and M. T. Kinnison. 2003. Contemporary evolution meets conservation biology. Trends in Evolution and Ecology 18:94–101.

Sussman, R. W., G. M. Green, and L. K. Sussman. 1994. Satellite imagery, human ecology, anthropology, and deforestation in Madagascar. Human Ecology 22(3):333–354.

Thayer, R. L. 2003. LifePlace: bioregional thought and practice. University of California Press. Berkeley.

Tober, J. A. 1989. Wildlife and the public interest: nonprofit organizations and federal wildlife policy. Praeger Publishers. Westport, CT.

Todd, K. 2001. Tinkering with Eden: a natural history of exotics in America. W. W. Norton and Co. New York.

Trubek, A. B. 2008. The taste of place: a cultural journey into terroir. University of California Press, Berkeley.

Van Dyke, F. 2003. Conservation biology: foundations, concepts, applications. McGraw Hill. New York.

Walker, B., and D. Salt. 2006. Resilience thinking: sustaining ecosystems and people in a changing world. Island Press. Washington, DC.

Warren, L. S. 2003. American environmental history. Blackwell Publishing. Oxford.

Weiner, J. 1994. The beak of the finch: a story of evolution in our time. Knopf. New York.

Weisman, A. 2007. The world without us. Picador Press. New York.

Wilcove, D. S. 1999. The condor's shadow: the loss and recovery of wildlife in America. W. H. Freeman. New York.

Wilcove, D. S. 2008. No way home: the decline of the world's great animal migrations. Island Press. Washington, DC.

Wilson, E. O. 2003. The future of life. Vintage. New York.

Wilson, E. O., and F. M. Peter. 1988. Biodiversity. National Academy Press. Washington, DC.

World Conservation Monitoring Center. 1992. Global biodiversity: status of the Earth's living resources. Chapman and Hall. London.

Worster, D. 1994. Nature's economy: a history of ecological ideas. Cambridge University Press. Cambridge.

Young, T. P. 2000. Restoration ecology and conservation biology. Biological Conservation 92:73–83.

INDEX